Radar Imaging in Challenging Scenarios from Smart and Flexible Platforms

Radar Imaging in Challenging Scenarios from Smart and Flexible Platforms

Special Issue Editors

Stefano Perna
Francesco Soldovieri
Moeness Amin

MDPI • Basel • Beijing • Wuhan • Barcelona • Belgrade

Special Issue Editors

Stefano Perna
Department of Engineering (DI),
Università degli Studi di
Napoli "Parthenope"
Italy

Francesco Soldovieri
Institute for Electromagnetic
Sensing of the Environment,
National Research
Council (CNR)
Italy

Moeness Amin
Center for Advanced
Communications,
Villanova University
USA

Editorial Office
MDPI
St. Alban-Anlage 66
4052 Basel, Switzerland

This is a reprint of articles from the Special Issue published online in the open access journal *Remote Sensing* (ISSN 2072-4292) from 2018 to 2020 (available at: https://www.mdpi.com/journal/remotesensing/special_issues/radarimaging).

For citation purposes, cite each article independently as indicated on the article page online and as indicated below:

LastName, A.A.; LastName, B.B.; LastName, C.C. Article Title. *Journal Name* **Year**, *Article Number*, Page Range.

ISBN 978-3-03936-469-5 (Pbk)
ISBN 978-3-03936-470-1 (PDF)

Contents

About the Special Issue Editors

Stefano Perna completed his Laurea Degree (summa cum laude) in Telecommunication Engineering and his PhD in Electronic and Telecommunication Engineering, both at the Università degli Studi di Napoli "Federico II", Naples, Italy, in 2001 and 2006, respectively. Since 2006, he has been with the Department of Engineering (DI) at the Università degli Studi di Napoli "Parthenope", Naples, where he is currently a researcher in electromagnetics and teaches the courses "Antennas" and "Electromagnetics". He is currently also an adjunct researcher at IREA-CNR, Naples. Since 2015, he has collaborated with the Argentinian National Council of Technical and Scientific Research (CONICET) on activities relevant to the focusing and processing of SAR data acquired by the airborne SARAT system. In 2016, he was a visiting professor at the Departamento de Teoría de la Señal y Comunicaciones of the Universitat Politècnica de Catalunya (UPC), Barcelona, Spain. He is an IEEE senior member. His main research interests are in the field of microwave remote sensing and electromagnetics: airborne SAR data modelling and processing, airborne differential SAR interferometry, modelling of electromagnetic scattering from natural surfaces, synthesis of antenna arrays; antenna characterization and measurement in anechoic and reverberating chambers. He is the co-author of about 100 papers published in international scientific journals or proceedings of international conferences in the field of electromagnetics and remote sensing.

Francesco Soldovieri is a research director at Institute for Electromagnetic Sensing of the Environment of CNR. He was General Chair of the International Workshop on Advanced Ground Penetrating Radar 2007 and General Co-Chair of the Ground Penetrating Radar Conference 2010. He was a member of the Editorial Board of IEEE-GRSL and now of IEEE-TCI and IEEE-TGRS, *Remote Sensing* (MDPI). He is Editor in Chief of *HERITAGE*, a MDPI journal devoted to cultural and natural heritage. He was the scientific coordinator of the FP7 projects ISTIMES and AMISS and the technical manager of the H2020 Project HERACLES. He was the President of the Division on Geosciences Instrumentation and Data Systems of the European Geosciences Union. His research interests include radar imaging, data processing for GPR, indoor surveillance, through-wall imaging, passive radars, integration of geophysical data, and radars for planetary exploration. He is the co-author of about 240 papers in national and international journals, and more than 350 conference proceedings.

Moeness Amin completed his DSc degree at the Faculty of Engineering, Cairo University in 1976, and his MSc at the University of Petroleum and Minerals in 1980. He then completed a PhD at the University of Colorado, Boulder, in 1984; all degrees are in Electrical Engineering. Since 1985, he has been with the Faculty of the Department of Electrical and Computer Engineering, Villanova University, Villanova, PA, USA, where he became the Director of the Center for Advanced Communications, College of Engineering, in 2002. Dr. Amin is a Fellow of the Institute of Electrical and Electronics Engineers (IEEE); Fellow of the International Society of Optical Engineering (SPIE); Fellow of the Institute of Engineering and Technology (IET); and a Fellow of the European Association for Signal Processing (EURASIP). He is the recipient of: the 2017 Fulbright Distinguished Chair in Advanced Science and Technology; the 2016 Alexander von Humboldt Research Award; the 2016 IET Achievement Medal; the 2014 IEEE Signal Processing Society Technical Achievement Award; the 2009 Technical Achievement Award from the European Association for Signal Processing; the

2015 IEEE Aerospace and Electronic Systems Society Warren D White Award for Excellence in Radar Engineering. He is also the recipient of the IEEE Third Millennium Medal. Dr. Amin has over 800 journal and conference publications in signal processing theory and applications, covering the areas of wireless communications, radar, sonar, satellite navigations, ultrasound, healthcare, and RFID. He has co-authored 23 book chapters and is the editor of three books titled, *Through the Wall Radar Imaging, Compressive Sensing for Urban Radar*, and *Radar for Indoor Monitoring*, all published by CRC Press, in 2011, 2014, 2017, respectively.

Editorial

Editorial for Special Issue "Radar Imaging in Challenging Scenarios from Smart and Flexible Platforms"

Stefano Perna [1,2,*], **Francesco Soldovieri** [2] and **Moeness Amin** [3]

[1] Department of Engineering (DI), Università degli Studi di Napoli "Parthenope", 80143 Napoli, Italy
[2] Institute for Remote Sensing of Environment (IREA), National Research Council (CNR), 80124 Napoli, Italy; soldovieri.f@irea.cnr.it
[3] Center for Advanced Communications, Villanova University, Villanova, PA 19085, USA; moeness.amin@villanova.edu
* Correspondence: perna@uniparthenope.it

Received: 15 April 2020; Accepted: 15 April 2020; Published: 17 April 2020

Abstract: Microwave radar imaging plays a key role in several civilian and defense applications, such as security, surveillance, diagnostics and monitoring in civil engineering and cultural heritage, environment observation, with particular emphasis on disasters and crisis management, where it is required to remotely sense the area of interest in a timely, safe and effective way. To address these constraints, a technological opportunity is offered by radar systems mounted onboard smart and flexible platforms, such as ground-based ones, airplanes, helicopters, drones, unmanned aerial and ground vehicles (UAV and UGV). For this reason, radar imaging based on data collected by such platforms is gaining interest in the remote sensing community. However, a full exploitation of smart and flexible radar systems requires the development and use of image formation techniques and reconstruction approaches able to exploit and properly deal with non-conventional data acquisition configurations. The other main issue is related to the need to operate in challenging environments, and still deliver high target detection, localization and tracking. These environments include through the wall imaging, rugged terrain and rough surface/subsurface. In these cases, one seeks mitigation of the adverse effects of clutter and multipath via the implementation of effective signal processing strategies and electromagnetic modeling.

This Special Issue (SI) is aimed at providing an overview of recent scientific and technological advances in the field of radar imaging from smart and flexible platforms, in terms of hardware, modeling and data processing.

The contributions of the SI can be generally classified into two groups.

The papers belonging to the first group [1–6] provide the description of the capabilities of newborn imaging radar systems designed to operate in challenging scenarios [1] or using smart and flexible aerial platforms, such as small airplanes [2], drones [3–5] or helicopters [6]. Overall, these contributions provide an interesting survey of the potential of lightweight and compact imaging radar sensors. The described systems cover a very wide range of the microwave spectrum, including the VHF band, up to the X-band. The papers under this group [5] provide a good survey of the radar hardware as well as the corresponding processing chain applied to the acquired data

The contributions belonging to the second group [7–10] are focused on the description of novel data processing techniques aimed at achieving accurate radar imaging under complex acquisition geometries, such as in the case of airborne Synthetic Aperture Radar (SAR) [6–8], or in challenging scenarios, as in the case of Forward-Looking Ground-Penetrating Radar (FL-GPR) [9] or Lunar Penetrating Radar (LPR) [10].

As for the papers belonging to the first group, in [1], a newborn Ultra Wideband (UWB) Multiple-Input Multiple-Output (MIMO) radar system exploiting the Stepped-Frequency Continuous-Wave (SFCW) technology to detect human targets beyond the obstacle, is presented. More specifically, the design, as well as manufacturing processes leading to the realization of the overall radar system, which also includes a novel miniaturized Vivaldi antenna with 0.5–2.5 GHz bandwidth, are described. The radar system is successfully used for through-wall imaging applications by exploiting a data-processing algorithm based on the Cross-Correlation Time Domain Back Projection (CC-TDBP) technique.

In [2–4], two newborn SAR systems mounted onboard aerial platforms are presented. In particular, in [2], the imaging and topographic capabilities of a novel Italian airborne X-band SAR system, named AXIS, are discussed. The system is based on the Frequency-Modulated Continuous-Wave (FMCW) technology and is equipped with a single-pass interferometric layout. In this work, the description of the developed radar system is given along with a quantitative assessment of the quality of the SLC (Single Look Complex) SAR images and the interferometric products achievable through the system.

In [3,4], a novel Brazilian drone-borne SAR system operating in three different frequency bands, namely the C-, L- and P-band, is presented. The system is capable of exploiting a single-pass interferometric configuration at C-band, and full-polarimetric configurations at the L- and P-band. In [3], the description of the system and a quantitative assessment of the results achieved by applying the Differential SAR Interferometry (DInSAR) technique to the L-band data is presented. The work in [4] is focused on an interesting precision farming application scenario enabled by the exploitation of the drone-borne SAR system. More specifically, a novel methodology for obtaining growth deficit maps with an accuracy down to 5 cm and a spatial resolution of 1 m is presented. The proposed methodology is based on the DInSAR technique.

Another light and compact imaging radar system mounted onboard a small Multicopter-Unmanned Aerial Vehicle (M-UAV) is presented in [5]. In this case, the radar operates with 1.7 GHz bandwidth centered at 3.95 GHz, and the flight positions are obtained through the Carrier-Phase Differential GPS (CDGPS) technique. In particular, the work describes the overall radar imaging system in terms of both hardware devices and data processing strategy. The system is validated by collecting and processing a dataset through a single flight track to provide focused images of on surface targets.

In [6], a helicopter-borne integrated Sounder/SAR system operating in the UHF and VHF frequency bands is described. More specifically, the Sounder operates at 165 MHz, whereas the full-polarimetric SAR could operate either at 450 MHz or at 860 MHz. The system is developed under the auspices of a contract between the Italian Space Agency (ASI) and different private and public Italian Research Institutes and Universities. In this work, the first results relevant to a set of Sounder and SAR data, acquired during a campaign conducted in 2018 over a desert area in Erfoud, Morocco, are presented.

As for the papers belonging to the second group, they address the processing of three kinds of imaging radar data, namely, airborne SAR [7,8], FL-GPR [9] and LPR [10] data. For airborne SAR processing, exploitation of small and flexible aerial platforms to mount the radar systems makes the issues related to motion errors (that is, the attitude and position instabilities of the platform during the acquisition) coupled to the topographic variations of the observed scene even more critical; therefore, ad-hoc data processing strategies capable to properly account for these problems are needed.

In [7], the spatial variations induced on airborne SAR images by the motion errors are decomposed into three main parts: range, azimuth and cross-coupling terms. The cross-coupling variations are then corrected by means of a polynomial phase filter, whereas the range and azimuth terms are removed through Stolt mapping.

In [8], an extended back-projection approach is proposed to take into account the topography variations during the airborne SAR image formation process. In particular, the algorithm applies a time–frequency rotation operation to pursue high accuracy, while reducing the computational burden, typically required by standard back-projection algorithms operating entirely in the time-domain.

The FL-GPR allows fast scanning of large areas for real-time target detection, unlike its ground-coupled or near-ground down-looking GPR (DL-GPR) counterparts. This capability, however, comes at the expense of energy backscattered from the illuminated targets and limited image spatial resolution. Furthermore, the rough ground surface generates clutter that may obscure the buried targets, rendering target detection very challenging. In this respect, the work in [9] presents an enhanced imaging procedure for the suppression of the rough surface clutter arising in FL-GPR applications. The procedure is based on a matched filtering formulation of microwave tomographic imaging enhanced by a coherence factor (CF) scheme for clutter suppression.

The work in [10] is framed in the context of the planetary exploration and deals with the Lunar Penetrating Radar mounted onboard the Yutu lunar rover to detect the lunar regolith and the shallower subsurface geologic structures of the Moon. In particular, it is aimed at improving the capability of identifying response signals caused by discrete reflectors (such as meteorites, basalt and debris) beneath the lunar surface. To this end, a compressive sensing (CS)-based approach is proposed to estimate the amplitudes and time delays of the radar signals from LPR data.

In conclusion, this informative Special Issue would not have been possible without the hard work of all authors and reviewers. We also would like to extend our sincere appreciation to the Editorial Office of Remote Sensing for their professional and excellent management work.

Author Contributions: The authors contribute equally to write this Editorial. All authors have read and agreed to the published version of the manuscript.

Conflicts of Interest: The authors declare no conflict of interest.

References

1. Hu, Z.; Zeng, Z.; Wang, K.; Feng, W.; Zhang, J.; Lu, Q.; Kang, X. Design and Analysis of a UWB MIMO Radar System with Miniaturized Vivaldi Antenna for Through-Wall Imaging. *Remote Sens.* **2019**, *11*, 1867. [CrossRef]

2. Esposito, C.; Natale, A.; Palmese, G.; Berardino, P.; Lanari, R.; Perna, S. On the Capabilities of the Italian Airborne FMCW AXIS InSAR System. *Remote Sens.* **2020**, *12*, 539. [CrossRef]

3. Luebeck, D.; Wimmer, C.F.; Moreira, L.; Alcântara, M.; Oré, G.A.; Góes, J.P.; Oliveira, L.; Teruel, B.S.; Bins, L.H.; Gabrielli, L.; et al. Drone-borne Differential SAR Interferometry. *Remote Sens.* **2020**, *12*, 778. [CrossRef]

4. Oré, G.; Alcântara, M.; Góes, J.; Oliveira, L.; Yepes, J.; Teruel, B.; Castro, V.; Bins, L.; Castro, F.; Luebeck, D.; et al. Crop Growth Monitoring with Drone-Borne DInSAR. *Remote Sens.* **2020**, *12*, 615. [CrossRef]

5. Catapano, I.; Gennarelli, G.; Ludeno, G.; Noviello, C.; Esposito, G.; Renga, A.; Fasano, G.; Soldovieri, F. Small Multicopter-UAV-Based Radar Imaging: Performance Assessment for a Single Flight Track. *Remote Sens.* **2020**, *12*, 774. [CrossRef]

6. Perna, S.; Alberti, G.; Berardino, P.; Bruzzone, L.; Califano, D.; Catapano, I.; Ciofaniello, L.; Donini, E.; Esposito, C.; Facchinetti, C.; et al. The ASI Integrated Sounder-SAR System Operating in the UHF-VHF Bands: First Results of the 2018 Helicopter-Borne Morocco Desert Campaign. *Remote Sens.* **2019**, *11*, 1845. [CrossRef]

7. Tang, S.; Zhang, L.; So, H. Focusing High Resolution Highly-Squinted Airborne SAR Data with Maneuvers. *Remote Sens.* **2018**, *10*, 862. [CrossRef]

8. Lin, C.; Tang, S.; Zhang, L.; Guo, P. Focusing High-Resolution Airborne SAR with Topography Variations Using an Extended BPA Based on a Time/Frequency Rotation Principle. *Remote Sens.* **2018**, *10*, 1275. [CrossRef]

9. Comite, D.; Ahmad, F.; Dogaru, T.; Amin, M. Coherence-Factor-Based Rough Surface Clutter Suppression for Forward-Looking GPR Imaging. *Remote Sens.* **2020**, *12*, 857. [CrossRef]

10. Wang, K.; Zeng, Z.; Zhang, L.; Xia, S.; Li, J. A Compressive Sensing-Based Approach to Reconstructing Regolith Structure from Lunar Penetrating Radar Data at the Chang'E-3 Landing Site. *Remote Sens.* **2018**, *10*, 1925. [CrossRef]

Article

Design and Analysis of a UWB MIMO Radar System with Miniaturized Vivaldi Antenna for Through-Wall Imaging

Zhipeng Hu [1], Zhaofa Zeng [1,*], Kun Wang [1,2], Weike Feng [3], Jianmin Zhang [1], Qi Lu [1] and Xiaoqian Kang [1]

[1] College of Geo-Exploration Science and Technology, Jilin University, Changchun 130026, China
[2] School of Mining and Safety Engineering, Shandong University of Science and Technology, Qingdao 266590, China
[3] Graduate School of Environmental Studies, Tohoku University, Sendai 980-8579, Japan
* Correspondence: zengzf@jlu.edu.cn

Received: 26 June 2019; Accepted: 7 August 2019; Published: 9 August 2019

Abstract: The ultra-wideband (UWB) multi-input multi-output (MIMO) radar technique is playing a more and more important role in the application of through-wall detection because of its high resolution, lower antenna requirements, and efficient data capturing ability. This paper develops a novel UWB MIMO radar system using a stepped-frequency continuous-wave (SFCW) signal, which is designed to detect human targets behind the regular brick and concrete wall. In order to balance high range resolution and wall-penetration depth, a novel miniaturized Vivaldi antenna with desired bandwidth of 0.5–2.5 GHz was designed, simulated, manufactured, and successfully used in through-wall imaging. To suppress the artifacts in the focused image and reduce the computing complexity, the cross-correlation-based time domain back projection (CC-TDBP) algorithm was developed. In addition, a through-wall imaging model was established, based on which the effects of the wall on the refraction of electromagnetic (EM) waves and the reduction of velocity are compensated. Finally, different experiments were conducted for multiple stationary targets utilizing the designed radar system, and the improved BP-based algorithms are applied to focus the targets behind the wall more accurately. The reconstructed two-dimensional (2D) images illustrate that the designed MIMO radar system can successfully detect and image human targets in the air and behind the wall.

Keywords: MIMO radar; through-wall imaging; ultrawideband signal; SFCW; sparse array; back projection algorithm; Vivaldi antenna

---

1. Introduction

Ultra-wideband (UWB) through-wall radar (TWR), as an emerging technology, is used to detect targets blocked by obstacles. It has wide application prospects in military and civil fields, such as urban combat, antiterrorism, vigilance, security inspection, disaster rescue, and so on [1–6]. According to the working mode, TWR can be divided into synthetic aperture radar (SAR) systems and MIMO radar systems [7–13]. Conventionally, to get a high azimuth resolution, the synthetic aperture method (SAR) is extensively used. The transceiver of a SAR system is sequentially sled on a rail to provide synthetic aperture scanning imaging. However, this method requires long data acquisition times and its azimuth resolution is limited by the length of the rail. In addition, the TWR of the SAR mode has the disadvantage of large scale and high cost [10–13]. In this case, the MIMO radar technology provides a new platform to solve the above problems [7–9,14–20].

The MIMO radar technology was proposed in 2003 and 2004 [14,15]. It uses multiple transmitting and receiving antennas and transmits orthogonal waveforms, which can expand the aperture of the actual array elements and confer a better spatial sampling ability. In addition, the sparse MIMO array design of switched antenna can also meet the requirements of high scanning speed and detection ability in the practical application of through-wall imaging. Therefore, the switched antenna array radar is often considered as the MIMO radar. In recent years, the MIMO radar system has received extensive attention of scholars. Amin et al. [1,4] proposed the 'co-array'. The azimuth resolution of the imaging is improved by expanding the aperture of the virtual array which is formed by the transceiver array. Zhuge et al. [18,19] studied the equivalence of a UWB MIMO array using the point spread function (PSF). It is pointed out that the equivalent array is the spatial convolution of the PSF of the transmitting array and the receiving array in the far field. Based on this, a linear array design was proposed. Feng et al. [20] developed a MIMO array-based radar system using a SFCW signal, which can effectively improve the data sampling speed, maintain the azimuth resolution, and reduce the hardware costs. However, due to the high frequency range of the antenna used, the penetration depth of the radar system is too shallow for through-wall imaging. Yılmaz et al. [6] designed and manufactured a uniform array radar system to detect and image targets behind a wall with a monostatic configuration controlled by a switcher. However, the effect of the wall on the refraction of EM waves, which produces a position shift of the targets behind the wall, was ignored.

Based on the advantages and limitations of the above research results, a UWB MIMO sparse array radar system with eight pairs of transmitting and receiving miniaturized Vivaldi antennas is studied in this paper and is mainly used for 2D imaging of multiple targets behind a wall. The UWB MIMO system combines UWB technology with MIMO technology. On one hand, UWB technology can improve range resolution [21,22]. On the other hand, the sparse topology of the MIMO system can obtain larger array aperture and smaller element spacing by using fewer antennas. Moreover, higher azimuth resolution and better main/side lobe control are obtained. Therefore, the MIMO radar system only needs a few physical elements to achieve the same imaging effect as SAR. In this way, the number of array elements and the system cost are greatly reduced, while the aperture length is maintained. In addition, the electronic switching mode used in the MIMO radar system can improve the data acquisition speed.

In the MIMO radar system, the antenna is of great importance as it radiates power to the wall and detects the signals of the targets. UWB signals are suitable for the radar system while the through-wall penetration losses for general wall materials increase with frequency. On the other hand, lower working frequency band for antennas can provide better through-wall penetration but this also implies a larger size of the antenna [23]. Hence, the antenna working frequency range of 0.5–2.5 GHz is considered to balance good resolution and strong penetration even for thick walls. In this paper, we apply Vivaldi antennas in a MIMO radar system for through-wall detection because of its many advantages, such as high gain, wide band, steady radiation patterns, simple structure, and simplicity of processing [24]. The low-frequency characteristic of antennas generally results in larger size. Inspired by [25,26], we use a slotted approach to change the surface current distribution to optimize performance throughout the band, expand the low frequency bandwidth, and reduce its size. Finally, the low-frequency UWB miniaturized Vivaldi antenna was designed and fabricated with dimensions of 250 mm × 200 mm × 1.6 mm, and a working band (the S11 is less than—10 dB) of 0.5–2.5 GHz, good end-fire radiation behavior, and acceptable gain. The Vivaldi antenna designed in this paper is one of the contributions to the imaging research of through-wall radar technology. The proposed antenna is simple, easy to fabricate, and is made of inexpensive FR4 substrate. The experiments show that the designed antenna has good through-wall penetration ability and resolution.

2D imaging is one of the most important signal processing steps for a through-wall MIMO radar system. The small imaging area is located in the near field of the radar. At the same time, EM waves produce reflections, attenuations, and velocity changes when they penetrate the wall. Therefore, the back projection (BP) imaging algorithm [9,20], which has no requirement for the array configuration, is one of the most practical through-wall imaging methods due to its convenience and robustness. However, reconstructed images of targets by the BP method may have some artifacts. In this paper, the cross-correlation-based time-domain BP algorithm (CC-TDBP) is applied to suppress artifacts [27,28]. Moreover, considering the effect of walls on the refraction and velocity of EM waves, an improved BP-based algorithm is proposed in Section 3. The experimental results validate that the designed MIMO radar system with the proposed BP-based algorithm can accurately reconstruct the human locations behind the wall.

The structure of the paper is organized as follows: In Section 2, the design process of the UWB MIMO array radar system is described. The imaging signal model and CC-TDBP algorithm are reviewed. The topology, array pattern of the designed MIMO array, and the 2D simulation imaging results are given. Moreover, the design methods of the miniaturized Vivaldi antenna are introduced. Finally, the prototype of the MIMO radar system is presented. In Section 3, the through-wall imaging model is constructed and the improved CC-TDBP algorithm, considering the effect of the wall, is illustrated. In Section 4, three experiments are conducted to assess the effectiveness of the developed MIMO radar system and imaging methods. To summarize, Sections 5 and 6 give discussion and conclusions, respectively.

2. UWB MIMO Radar System

The block diagram of the designed UWB MIMO radar system is shown in Figure 1. The system includes: a PC, a two-port VNA (Agilent N9925A, working from 30 kHz to 9 GHz), two 1–8 radio frequency (RF) switchers with one microcontroller, a 16-Vivaldi-antenna MIMO array with about 1.1 m aperture. As the master controller, the PC connects the VNA and microcontroller. A set of RF cables and two switchers connect eight transmitting antennas and eight receiving antennas to the VNA. The eight transmitting antennas are fixed in the middle of the array, and the eight receiving antennas are separately fixed on the upper and lower sides of the array (shown in Figure 1).

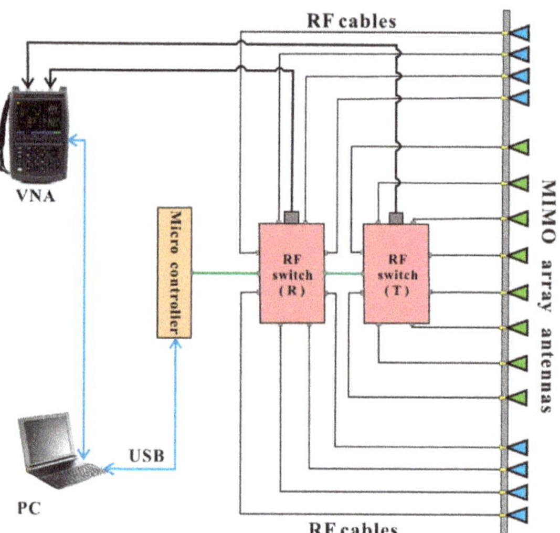

Figure 1. Block diagram of the designed ultra-wideband multi-input multi-output (UWB MIMO) radar system.

The working principle of the designed radar system is briefly explained here: the VNA is responsible for sending and receiving SFCW signals. The PC is not only used to save and process the received data, but also to remotely operate the microcontroller and VNA. During one period of every S12 measurement by the VNA, only one pair of transmitting and receiving antennas collects data, corresponding to two paths of switcher system staying open and all other paths turning off. By switching on and off the switchers to control the data acquisition, the total amount of data obtained in one sampling period is 64 channels.

2.1. SFCW Signal Model and Conventional BP Algorithm

For the SFCW based radar system, the range resolution is determined by the frequency bandwidth. The UWB signal can be used to obtain higher range resolution with a low hardware requirement. Therefore, we utilize the UWB SFCW signal model in this paper.

Figure 2 shows a linear MIMO sparse array with an SFCW signal, which contains M transmitters and N receivers. Assuming the m-th transmitter and n-th receiver are located at $(x_m, 0)$ and $(x_n, 0)$, where $m = 1, 2, \ldots, M$ and $n = 1, 2, \ldots, N$. Given a point target located at (x_0, y_0), the average distance between the target and the l-th pair of transmitter and receiver (l-th sampling point) is given by

$$R_l = (R_m + R_n)/2 = \left[\sqrt{(x_m - x_0)^2 + y_0^2} + \sqrt{(x_n - x_0)^2 + y_0^2} \right]\Big/2 \tag{1}$$

where $l = (m - 1)N + n = 1, 2, \ldots, MN$.

Then, we assume that the whole imaging scene is discretized by I grids in the x axis and J grids in the y axis. There are a set of targets and each one is located at a given grid in the x-y coordinate. For the l-th sampling point and the q-th frequency $f_q = f_0 + (q - 1)\Delta f$ the demodulated received signal can be expressed as

$$S(l, q) = \sum_{i=1}^{I} \sum_{j=1}^{J} \alpha(x_i, y_j) \exp(-j2k_q R_l(x_i, y_j)) + n(l, q) \tag{2}$$

where $k_q = 2\pi f_q/c$ is the wavenumber, $R_l(x_i, y_i)$ is the average distance between the grid at (x_i, y_j) and the l-th sampling point; f_0 is the start frequency, Δf is the frequency step, $\alpha(x_i, y_j)$ is the reflection coefficient of target, c is the velocity of light, and $n(l, q)$ is the noise.

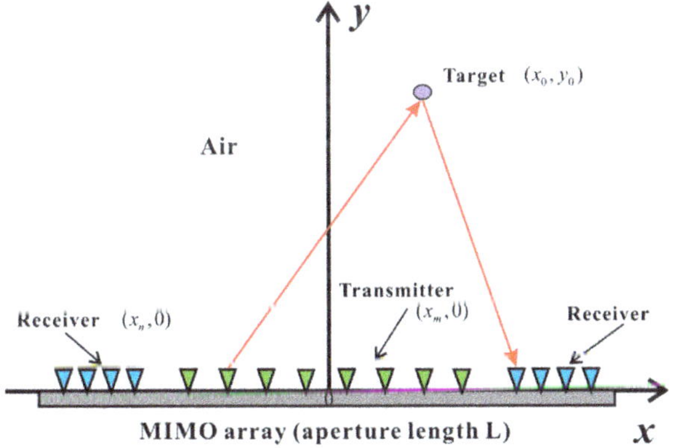

Figure 2. The signal model of the UWB SFCW radar with a MIMO array.

The traditional BP algorithm is applied in the case of free space or uniform medium, and the target echo delay directly corresponds to the linear distance between the target and the antenna. One of the typical BP methods is the frequency-domain back projection (FDBP) method. It coherently sums the received signals from all the frequencies and sampling points to estimate the reflection coefficients of targets. With the received signal $S(l, q)$, the reflection coefficient of a point (x_i, y_j) is estimated by

$$\widetilde{\alpha}(x_i, y_j) = \frac{1}{QL} \sum_{q=1}^{Q} \sum_{l=1}^{L} S(l,q) \exp(+\mathrm{j}2k_q R_l(x_i, y_j)) \tag{3}$$

In practice, the time-domain implementation of BP method (TDBP) [9,20] is most commonly used in the highly suboptimal aperture length case for its simplicity. It can significantly save computing time. The formulation of TDBP can be written as

$$\widetilde{\alpha}(x_i, y_j) = \frac{1}{L} \sum_{l=1}^{L} S_t(l, 2R_l(x_i, y_j)/c) \tag{4}$$

where

$$S_t(l, 2R_l(x_i, y_j)/c) = \frac{1}{Q} \sum_{q=1}^{Q} S(l,q) \exp(+\mathrm{j}2k_q R_l(x_i, y_j)) \tag{5}$$

Equations (4) and (5) are the range compressed signal of the l-th sampling point, which can be easily obtained by the inverse fast Fourier transformation (IFFT) and interpolation process.

Although TDBP imaging methods are simple and convenient to implement, their imaging results may have high-level artifacts. Strong artifacts make weak targets undetectable and artifacts of many targets produce spurious peaks, leading to negative effects on 2D imaging. In order to effectively suppress artifacts, the cross-correlation-based TDBP (CC-TDBP) algorithm is applied [27,28], which can be expressed as

$$\widetilde{\alpha}(x_i, y_j) = \sum_{l_1=1}^{MN-1} \sum_{l_2=l_1+1}^{MN} S_t(l_1, \tau(l_1, x_i, y_j)) S_t(l_2, \tau(l_2, x_i, y_j)) \tag{6}$$

where $\tau(l_1, x_i, y_j) = 2R_{l_1}(x_i, y_j)/c$ and $\tau(l_2, x_i, y_j) = 2R_{l_2}(x_i, y_j)/c$.

2.2. MIMO Array Topology Design

The design of the MIMO array is related to the complexity and cost of the radar system and directly affects the imaging quality. In designing MIMO array configurations, sparse array designs are often used to reduce system complexity and the number of antenna elements. In order to simplify the design of MIMO arrays, we first discuss the design factors from the following aspects.

Firstly, the azimuth resolution is determined by the aperture length of the array. Therefore, the azimuth resolution can be improved by properly increasing the aperture length of the array.

Secondly, after increasing the length of the array aperture, in order to satisfy the ideal imaging performance, the spacing and total number of antenna elements will also change. Specifically, due to the sparsity of the MIMO array, grating lobes and sidelobes may be generated in the imaging results, which seriously affect the quality of radar imaging. According to Nyquist sampling theory, the element spacing should be less than half of the radar wavelength in order to prevent unwanted grating lobes. On the other hand, with the decrease of the element spacing and the increase of the number of antennas, the direct coupling among the antennas will become stronger and high-level sidelobes will occur [29,30]. Therefore, considering the above factors, the antennas should keep a certain distance and the number of antennas should not be too large.

Finally, the imaging performance of UWB MIMO radar is related not only to the array configuration and number of elements, but also to the form of transmitted signals and application scenes. In this paper, a UWB SFCW signal is used to detect the short-range targets behind the wall. Therefore, it is necessary to consider the applicability of near-field conditions in through-wall detection. Generally, the longitudinal distance from the target to the MIMO array is much larger than the lateral distance between the target and the array elements. According to the verification results of [20,29–32], the focused SAR imaging and PSF results of MIMO sparse arrays and monostatic uniform arrays have almost the same performance in near-field conditions. Therefore, as for the MIMO through-wall radar system in this paper, the virtual array theory is applied to simplify the design and analysis of the MIMO sparse array.

For a bistatic MIMO sparse array which contains M transmitters and N receivers, according to the theory of displaced phase center (DPC) approximation [18,33], two paths of the m-th transmitter at the x_m-th position and the n-th receiver at x_n-th position are switched on, while all the other paths are switched off. This is equivalent (in far field) to transmitting and receiving with one single 'virtual' antenna in the median position $(x_m + x_n)/2$ of the axis. The midpoints of each transmit and receive element are regarded as virtual elements for a linear equivalent monostatic array. Hence, for each combination of the m-th and n-th antennas, a specific pattern along the median axis is defined. As a result, we can first obtain an equivalent uniform array with MN ($MN = M{\times}N$) monostatic transceivers, which has expected resolution and low-level grating/side lobes. Then, corresponding to the reverse thinking process mentioned above, the MIMO sparse array can be designed by factorization. In this case, a synthetic aperture imaging algorithm, such as the BP-based imaging algorithm, can be directly used to image the targets.

Based on the above analysis, the design steps of the MIMO array, which are illustrated in Figure 3, are as follows: Firstly, we begin with designing a linear monostatic uniform equivalent array. The effective aperture size L relative to the wavelength of the center frequency is determined by the required cross-range resolution $\delta_a = \lambda R_0/2L$, where λ is wavelength, L is equivalent uniform array length, and R_0 represents the potential distance of the target. Then, the required ideal sidelobe level (SL) is limited by the lower bound $SL = -20\log(K)$ (dB) [21], which helps to derive the minimal number of the virtual elements $K = MN$ required within the effective aperture, where K refers to the channels of the radar array. The next step is to determine the configuration of MIMO array, which is the factorization of a desired equivalent uniform array into topologies of transmitting and receiving MIMO arrays. The number of transmitting and receiving antennas can be obtained from the relation $K = M \times N$ (the total number of antenna elements is $Num = M + N$). The total number of antennas is minimized by selecting an equal or as close as possible number of elements in both transmitting and receiving arrays. For example, the number of virtual elements in the paper is 64; therefore, four different combinations of transmitting and receiving antenna element numbers can be obtained, which are: 1×64, 2×32, 4×16, and 8×8. The result with the minimal number of antennas is eight transmitters and eight receivers (8×8). At last, based on the polynomial factorization (PF) method in our previous research work [20], a MIMO array with eight pairs of transmitters and receivers was achieved, which satisfies the desired resolution and sidelobe level. The length of aperture is about 1.1 m. The operational center frequency is 1.5 GHz. The array possesses 6.8 cm range resolution and 5.8 rad azimuth resolution, with an equivalent aperture length of 1 m.

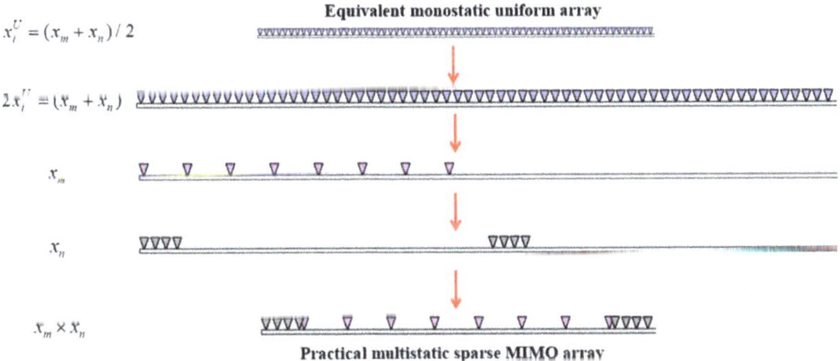

Figure 3. Schematic diagram of the polynomial factorization (PF) method.

The designed MIMO array with 16 antennas can be used to collect 64 channels of raw data by switching on and off all antennas in a sampling period. This array design method can bring about great reduction in the total number of antenna elements and improve aperture efficiency while maintaining the aperture size. The topology of the designed MIMO array is shown in Figure 4. To assess the capacity of the designed MIMO array, given a point target located at (0,5), the point spread function (PSF) [20] of a sparse array in the azimuth direction is calculated, as given in Figure 5. We can see that the designed MIMO array can achieve low sidelobe level.

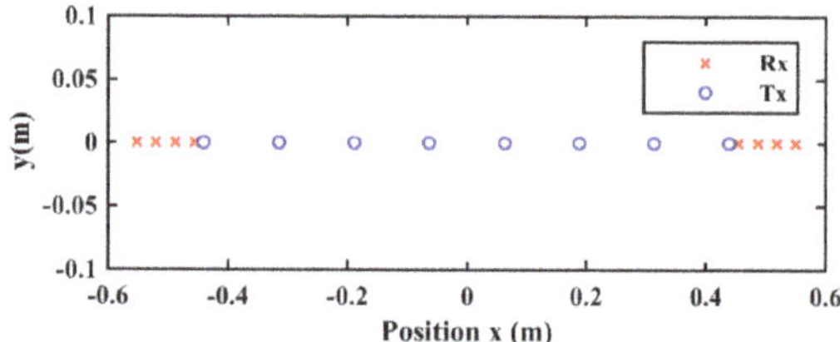

Figure 4. Topology of the designed MIMO sparse array.

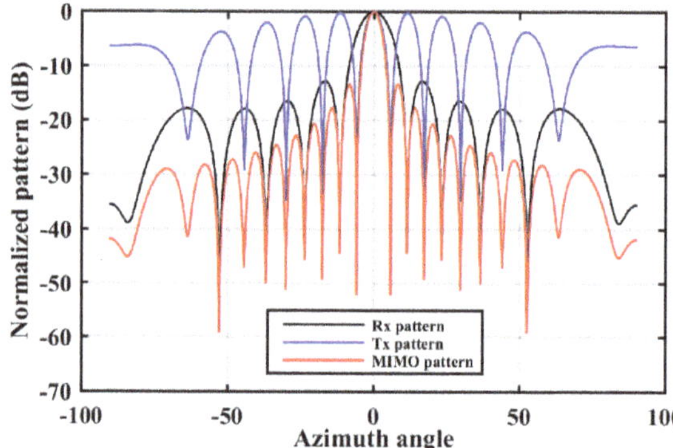

Figure 5. Point spread function (PSF) of the designed MIMO sparse array in the azimuth direction.

In order to assess the imaging quality of the above array design results and imaging methods, simple numerical simulations were carried out. The simulation parameters (Table 1) are consistent with the setting parameters of the actual system. Three targets located at different positions, which are $(-1,3)$, $(0,1)$ and $(1.5,5)$, were simulated. Figure 6 shows the 2D imaging performance of the TDBP algorithm and the CC-TDBP algorithm from the designed layout of the MIMO array, respectively. It can be seen that the targets can be well focused. The artifacts in Figure 6b are suppressed compared with those in Figure 6a.

Table 1. System parameters.

Parameters	Value
Start frequency	0.4 GHz
Stop frequency	2.6 GHz
Number of frequencies	256
Number of transmitters/receivers	8/8
Range resolution	0.068 m
Azimuth resolution	5.8 rad
Maximum range	17.4 m
Length of aperture	1.1 m

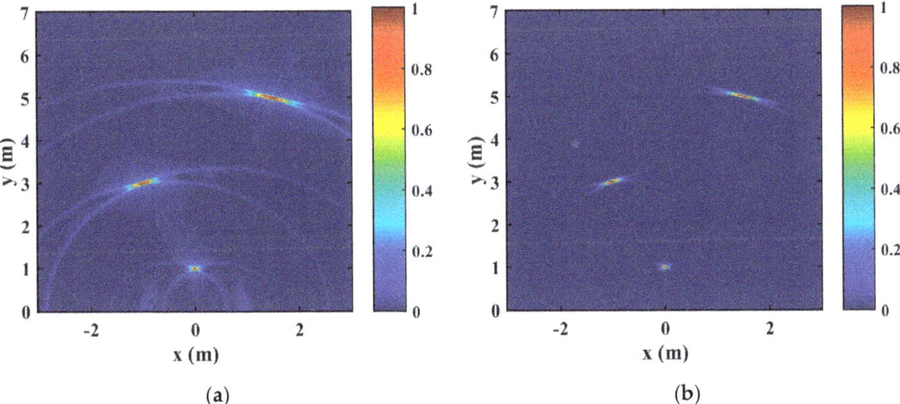

(a)   (b)

Figure 6. Simulations for the MIMO sparse array by: (**a**) the time domain back projection (TDBP) algorithm; (**b**) the cross-correlation-based time domain back projection (CC-TDBP) algorithm.

2.3. Low-Frequency UWB Miniaturized Vivaldi Antenna

The Vivaldi antenna is a kind of UWB tapered slot antenna. It has the advantages of wide bandwidth, symmetrical pattern, stable gain, simple structure, low profile, and easy integration [23,24]. Generally speaking, the size of antenna is determined by its working wavelength, and the wavelength λ can be obtained from the relationship ($\lambda = c/f$) between the wave velocity c and frequency f in free space. Therefore, it can be clearly seen from the above relationship that the lower the operating frequency of the antenna, the larger the size of the antenna required. According to the needs of some specific applications, such as a portable MIMO antenna array for TWR imaging in this paper, we plan to use the UWB Vivaldi antenna with low frequency characteristics, so it is necessary to study the miniaturization of Vivaldi antenna. Through a series of studies and discussions [23–26,34–36], it can be concluded that most antenna sizes are reduced but many aspects of performance will be affected, such as narrowing of the working frequency band and decreasing of gain. As a result, the research on antenna miniaturization is to ensure that the performance of the antenna is minimally changed.

A miniaturization technique commonly used with Vivaldi antennas is to etch shorting slots with different shapes on the nonradiative sides of the metal patch [25,26,34,35]. Firstly, according to the structural analysis of the conventional tapered slot Vivaldi antenna, the radiation mechanism consists of current flowing effectively along the edge of the mid-tapered slot line and radiating outward. However, surface current and edge current also exist in the nonradiating metal areas on both sides of the antenna, which will affect the antenna gain and directivity. Then, some regular rectangle slots are etched on the nonradiative sides and the direction of the rectangle slot is perpendicular to the direction of the current. In this case, the flow path of the current is significantly increased. The rectangle slot edge structure on both sides of the antenna has infinite impedance for surface waves. It can effectively suppress the radiation of surface waves at the nonradiation sides so that the current flows along the mid-tapered slot line. As a result, the operating frequency of the antenna can be lowered without changing the size of the antenna and the operating mode.

Above all, the final dimensions of optimized Vivaldi antenna structure in the results simulated by HFSS 15.0 are 250 mm × 200 mm × 1.6 mm, which is designed to work across the desired bandwidth of 0.5–2.5 GHz. Figure 7 shows the final design and fabrication result of the Vivaldi antenna. It is made of the low-cost FR4 substrate for which the dielectric constant is about 4.6. The feeding of the antenna is achieved by a subminiature version A (SMA) connector. Figure 8 plots both the E-plane and H-plane radiation patterns of the designed antenna at 0.5 GHz, 1.5 GHz, and 2.5 GHz, which are simulated using HFSS 15.0. It can be seen that the proposed Vivaldi antenna has good end fire radiation behavior and acceptable directivity in bandwidth of 0.5–2.5 GHz. Then, the antenna is measured by vector network analyzer (VNA) FieldFox N9925A. A comparison between the measured S11 with shorting slots and simulated S11 with and without shorting slots for the antenna is shown in Figure 9a. The −10 dB frequency bandwidth of 0.5–2.5 GHz is apparent. The measured result agrees well with the designed wideband except for the slight degradation of the low frequency starting band and the band near 2.0 GHz, where the S11 parameter rises to be a slightly higher than −10 dB. To see the tailing effect of the antennas for the time domain signal, we placed two identical Vivaldi antennas at the same altitude in parallel and face to face in the air, and then collected data once in the VNA's S12 measurement mode. After simple IFFT processing, the single channel time domain wavelet is obtained. Figure 9b shows the one-channel wavelet pulse signal of the designed antenna in the time domain. Figure 10 shows the simulated total gain in the direction of the maximum radiation as a function of frequency for the designed Vivaldi antenna. We can see that the gain of the designed antenna increases as the frequency gets higher and it varies between 5.1 dB and 10.1 dB across the desired 0.5 to 2.5 GHz bandwidth. Compared with the UWB antennas for similar applications in [23], higher values of gain are provided in this paper (5–10 dB versus 2–7 dB).

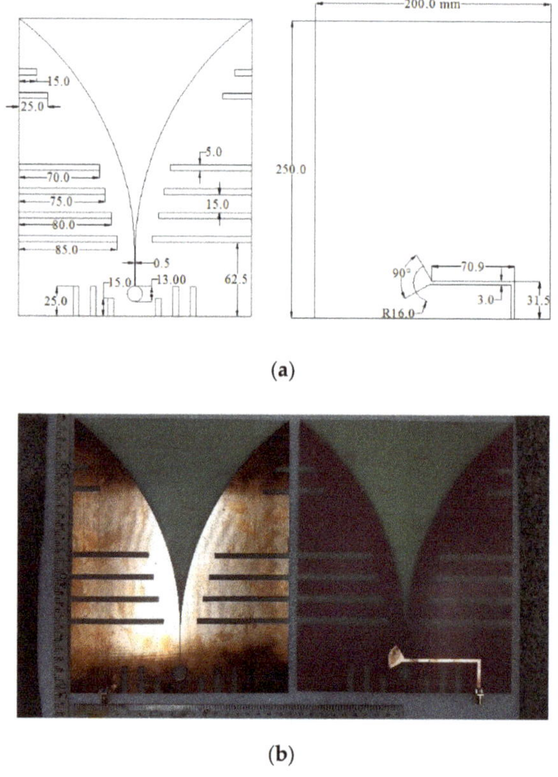

(a)

(b)

Figure 7. (a) Configurations of the designed Vivaldi antenna; (b) Top and bottom view of the manufactured Vivaldi antenna.

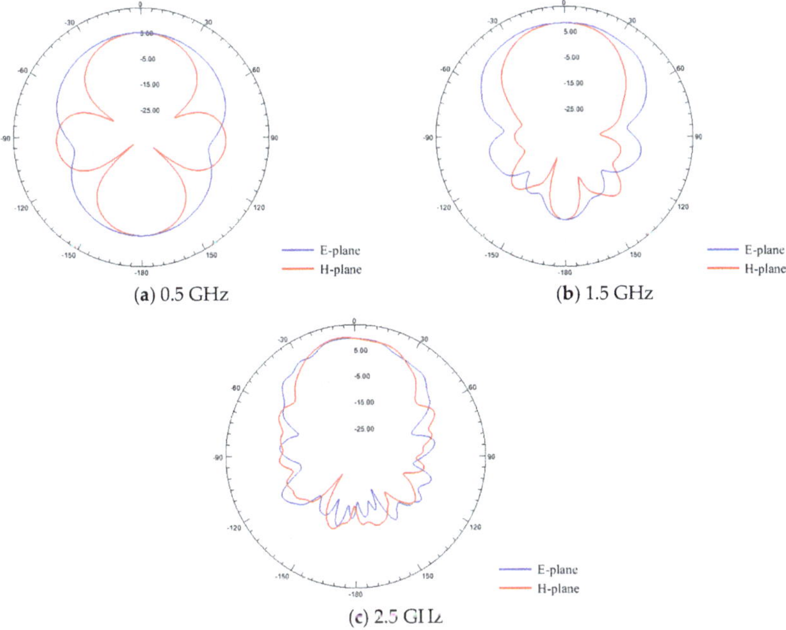

Figure 8. Simulated E-plane and H-plane radiation patterns of the designed Vivaldi antenna. (**a**) 0.5 GHz, (**b**) 1.5 GHz, and (**c**) 2.5 GHz.

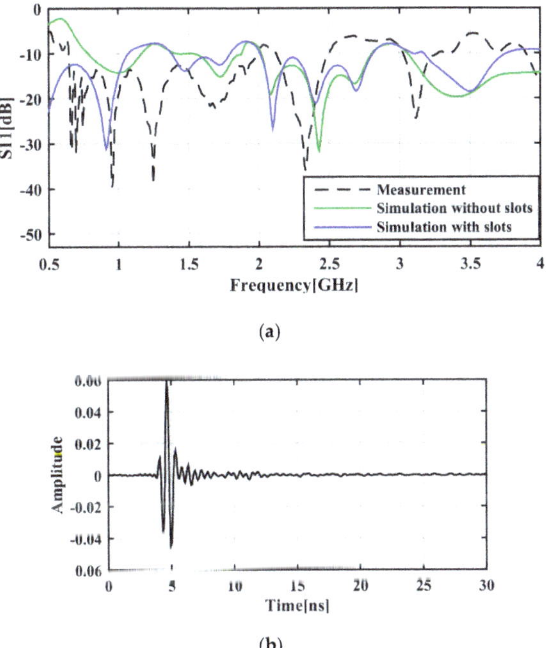

Figure 9. (**a**) The measured S11 of designed Vivaldi antenna with shorting slots and the simulated S11 of the proposed Vivaldi antenna with and without shorting slots; (**b**) The measured result of S12 in time domain of the designed Vivaldi antenna with slots.

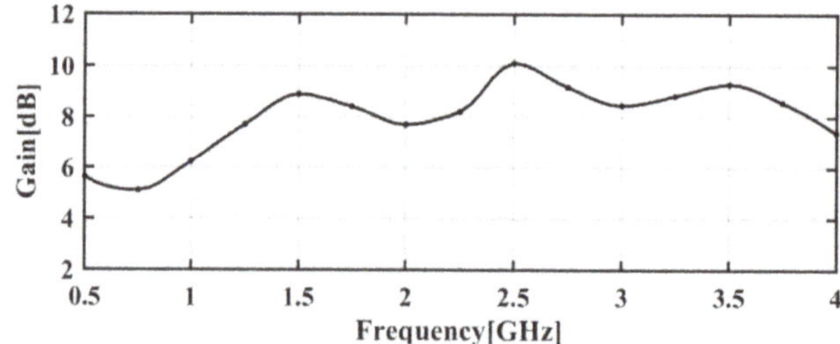

Figure 10. Total gain as a function of frequency of the designed Vivaldi antenna.

2.4. Radar Prototyping Result

According to the above antenna design method, 16 identical Vivaldi antennas were fabricated. Then, the array was assembled by the proposed design method. All antenna elements were connected to each terminal of the RF switch through SMA connectors and RF cables. For the switch at the center frequency of 1.5 GHz, the insertion loss was about 1.8 dB, and the isolation was greater than 48 dB. The control and the synchronization of VNA and the RF switch were realized by a microcontroller with the help of the PC. Finally, a MIMO radar system for through-wall detection was manufactured. Corresponding to the radar design block diagram in Figure 1, Figure 11 displays the photography of the designed MIMO radar system.

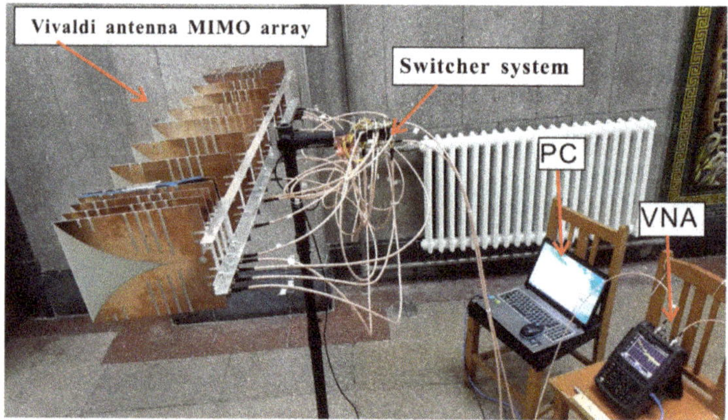

Figure 11. Photograph of designed MIMO radar prototype.

The time delay and phase difference among the 64 (8 × 8) channels of the designed MIMO array radar system were compensated by using mechanical calibration of the VNA. To avoid further calibration procedures, all the paths between each antenna and the VNA were of the same EM length. Therefore, only one path required calibration. The designed parameters of the TWR system are given in Table 1.

3. Through-Wall Imaging Method with the Improved BP-Based Algorithm

The traditional cross-correlation based TDBP algorithm is simple and practical in free space. However, when considering the existence of a concrete wall with 22.5 cm thickness, the imaging results focused directly by the BP imaging algorithm will cause errors between the imaged positions and the true positions.

Generally speaking, the influence of the wall on the location of target imaging is mainly due to the following two reasons: firstly, the slow propagation of EM waves in the wall produce additional time delay. This causes the targets' positions in the imaging results to be farther from than their actual positions. Secondly, the refraction and scattering of EM waves inside and outside the wall mean that the original propagation path is not a straight line. These errors are mainly affected by the relative dielectric constant of the wall. The above two points are briefly illustrated in Figure 12. In this paper, we propose an improved TDBP algorithm to reconstruct target locations behind the wall from MIMO sparse array radar data.

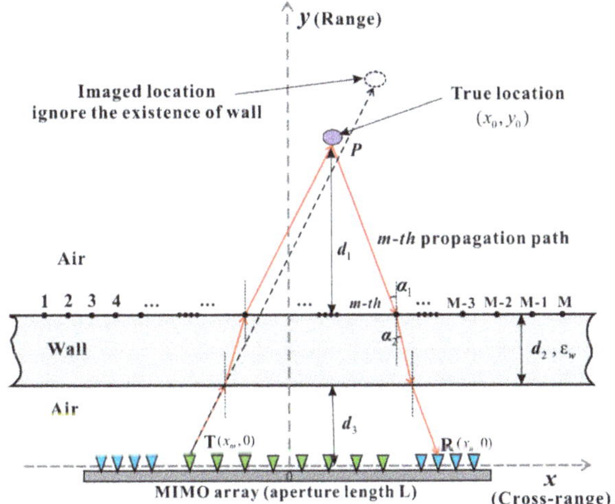

Figure 12. Imaging geometry for through-wall radar (TWR) with the designed MIMO array.

In the TDBP algorithm, the travel time for the wave field to propagate from the transmitter to the target and scatter back to the receiver must be known. In [37], we used ray tracing methods to efficiently calculate travel time in the TDBP algorithm for SIMO radar data. Assume that there are M paths passing through the wall from one antenna. The point p in the imaging area is a pixel on the m-th path. The incident angle is α_1. The refraction angle is α_2. The average propagation velocity of an EM wave inside the wall is v_w. The relative dielectric constant of the wall is ε_w. The thickness of the wall is d_2. The distance from the equivalent phase center of MIMO antenna array to the surface of the wall is d_3. The distance from the point p to the wall is d_1. According to Snell's law, we can obtain

$$\frac{\sin \alpha_1}{\sin \alpha_2} = \frac{c}{v_w} \tag{7}$$

where $v_w = c \sqrt{1/\varepsilon_w}$.

The travel time from the one antenna to the point p is

$$T_p = \left(\frac{d_1 + d_3}{\cos \alpha_1} + \frac{d_2 \sqrt{\varepsilon_w}}{\cos \alpha_2} \right) \frac{1}{c} \tag{8}$$

Then, the one-way travel time of all M propagation paths can be calculated in the same way, and the travel time of any point behind the wall can be obtained by interpolation from the M paths. The one-way travel time of any antenna can be computed similarly. Considering that there may be a large number of antennas in a MIMO array (16 in this paper), we designed a parallel algorithm on the MATLAB platform to improve the computational efficiency of one-way travel time (Figure 13, Part I). Assume the one-way travel time in the imaging area from the transmitter T_{x1} is T_{t1} and the receiver R_{x1} is T_{r1}. For a pair of transmitting and receiving antennas, T_{x1} and R_{x1}, the whole travel time behind the wall can be obtained by $T_1 = T_{r1} + T_{t1}$. According to the designed sparse MIMO array, we can obtain the whole travel time from all transmitters and receivers (at the pair of T_{x1}-R_{x2}, R_{x3}, ... R_{x8} to $T_{x8} - R_{x8}$) to the imaging area behind the wall (Figure 13, Part II).

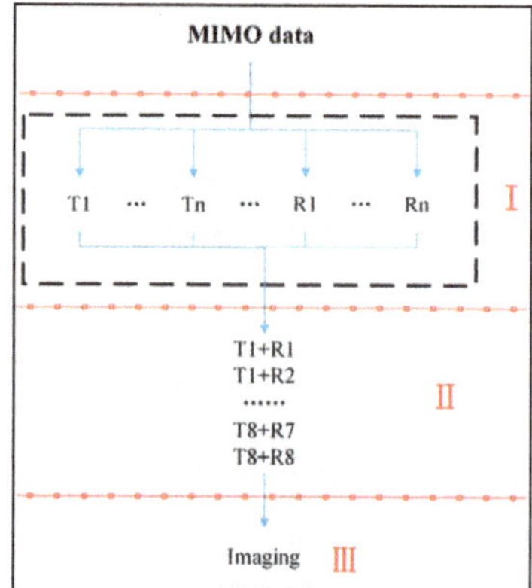

Figure 13. Flow chart of the improved TDBP algorithm. Part I: One-way travel time calculation based on the parallel algorithm; Part II: the whole travel time for the MIMO sparse array; Part III: TDBP imaging.

Finally, the TDBP algorithm can be directly applied to focus the data received from the MIMO antennas and reconstruct the target location (Figure 13, Part III). In this paper, $M = 501$ ray paths for every antenna are constructed, and the imaging range does not include the wall.

4. Experiment Results

In this section, three experiments are conducted to illustrate the performance and effectiveness of the designed MIMO radar system for 2D imaging. All parameters used in the experiments are the same as given in Table 1. The frequency range B of the radar system is 0.4–2.6 GHz, which results in the range resolution of $\delta_r = c/(2B) = 6.8$ cm. The number of frequencies is 256, which causes the unambiguous range to be 17.4 m. The S12 measurement mode for the Agilent N9925a VNA is selected. The imaging method is mainly based on the BP algorithm. In order to suppress the artifacts, the CC-TDBP method is also applied for comparison with the traditional TDBP algorithm. To make the target imaging clearer, background subtraction and direct coupling subtraction are also conducted for the real experimental data before applying the imaging algorithm. The direct coupling wave signal is obtained by collecting data while facing the MIMO array towards the open sky in an open area. Therefore, the received signal only contains the direct coupling component between the antenna elements. In general, the direct coupling data of one measurement result can be directly subtracted and applied to most 2D imaging of experimental scenes. The method of obtaining the background component is to measure the data without the real targets (human or corner reflector) in the experimental scene. Then, we can simply subtract the saved signal containing the background when we do the real measurement. This background removal method can eliminate the reflections of the wall, other stationary clutter, and the direct coupling between the antenna elements. It is relatively complicated and may be limited by the actual imaging situation. Because the background data are different for each specific imaging scene, the background data of each new experimental scene needs to be measured.

4.1. Preliminary Experimental Tests

In order to test the actual target imaging ability of the designed MIMO radar system, the experiments were preliminarily carried out in the air. In the first experiment, two small dihedral corner reflectors (DCR) were used as the targets in an open space. The positions of the two dihedral corner reflectors were approximately about 3 m and 5 m away from the equivalent phase-center of the radar system. The experimental setup is shown in Figure 14 and the imaging results are shown in Figure 15. It can be observed that the two targets can be focused in 2D imaging. Compared to the results obtained by background removal (Figure 15c,d), it is obvious that the method of removing the presampled direct coupling (Figure 15a,b) is simple and can roughly distinguish the positions of the DCR targets. However, due to the influence of clutter signals, such as ground reflection, there are more noise points and artifacts in the direct coupling removed images. It is worth noting that after utilizing the CC-TDBP algorithm, the imaging results can further suppress the artifacts and clearly distinguish the exact locations of the DCR targets, no matter the methods of removing direct coupling or background, which are shown in Figure 15) and Figure 15d, respectively. In conclusion, the CC-TDBP algorithm can significantly improve image quality and suppress artifacts.

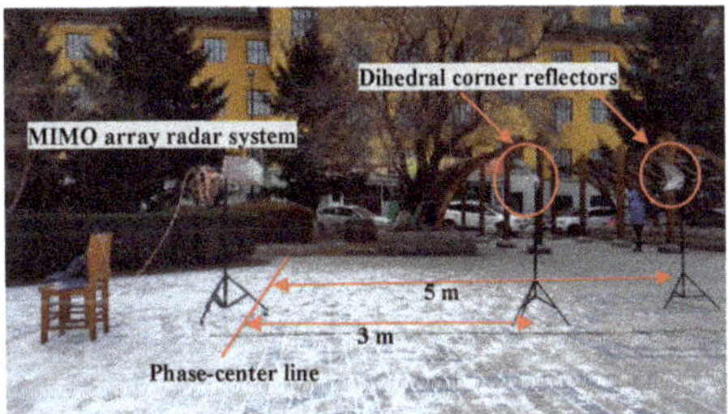

Figure 14. Experiment setup for dihedral corner reflector imaging.

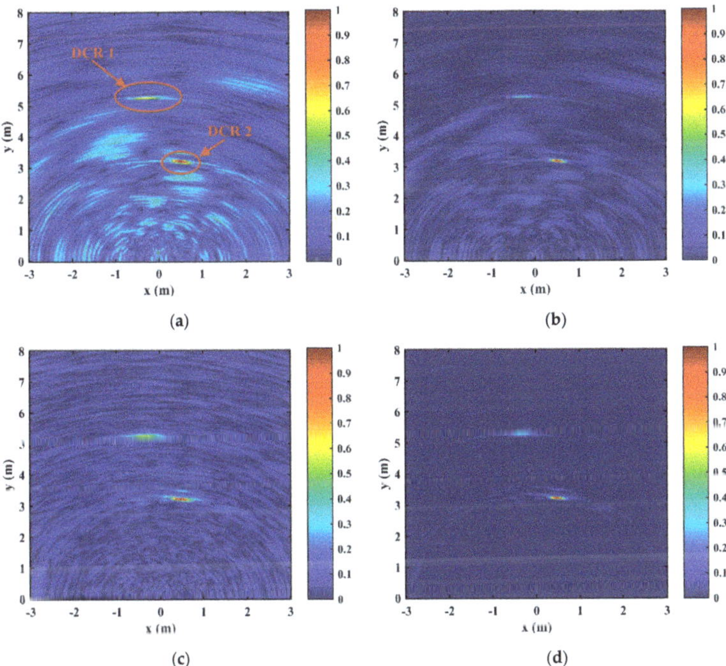

Figure 15. Imaging results of two corner reflectors by: (**a**) the TDBP algorithm with direct coupling removal; (**b**) the CC-TDBP algorithm with direct coupling removal; (**c**) the TDBP algorithm with background removal; (**d**) the CC-TDBP algorithm with background removal.

In the second experiment, to validate the human imaging performance of the designed radar system, two humans acted as targets in the near range, as shown in Figure 16. In this case, only the CC-TDBP algorithm was used, and the imaging results are shown in Figure 17. We can see that the two human targets marked by the red circle are clearly focused. In this preliminary experiment, even if the direct wave is removed (Figure 17a), the designed MIMO radar system can clearly detect the human targets in the air.

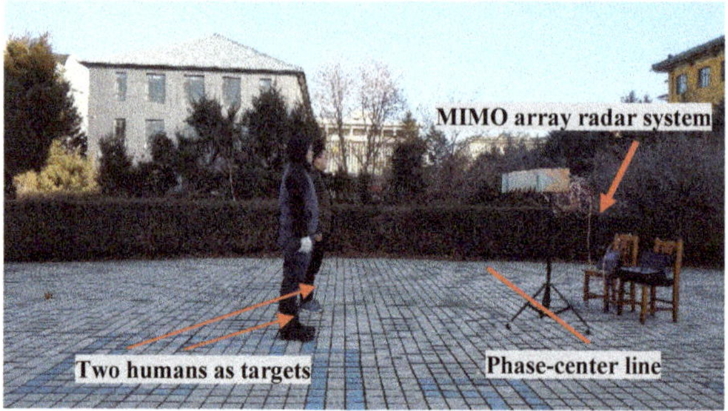

Figure 16. Experiment setup for human imaging.

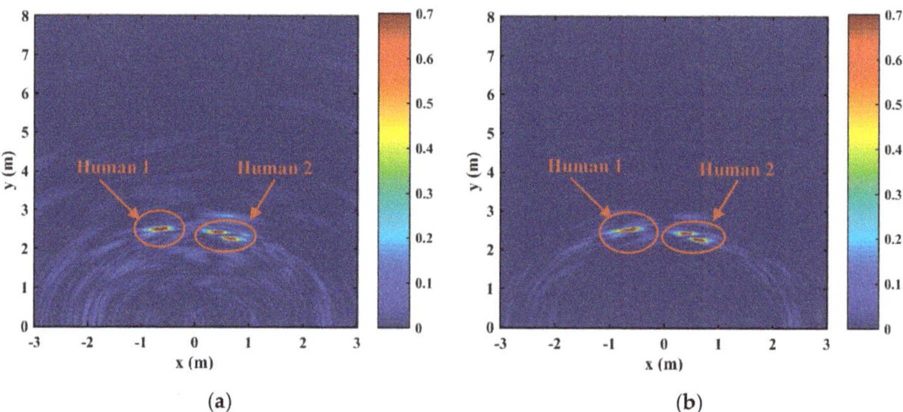

Figure 17. Imaging results obtained by the CC-TDBP algorithm with: (**a**) direct coupling removal; (**b**) background removal.

4.2. Through-Wall Experiment Results

In the third experiment, to assess the through-wall performance of the designed MIMO radar system and the proposed improved BP algorithms, the scene consisted of two humans hidden behind a solid wall with thickness of 22.5 cm. The walls are made of brick and concrete. After preliminary measurement (a former reflection test on the wall was conducted, which estimated that the speed of electromagnetic waves is 0.137 m/ns, gives a dielectric constant of 4.795 for the wall), the relative dielectric constant of the wall is about 4.8. The experimental setup is shown in Figure 18. The MIMO array was placed parallel along the azimuth direction on the side of the wall. The equivalent phase center line of the array was measured about 15 cm away from the surface of the wall. The original point of the coordinate system of the imaging region was the midpoint of the equivalent phase center line.

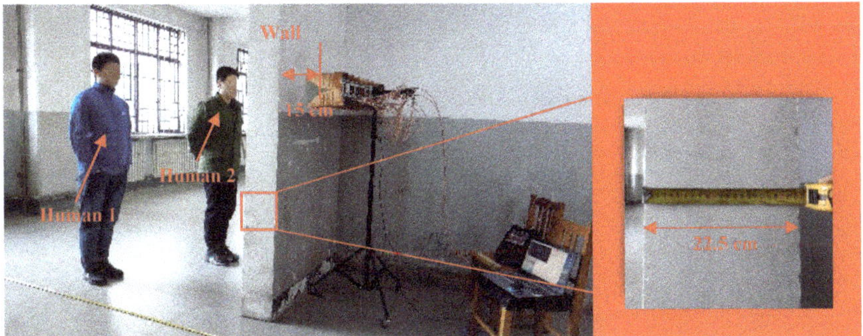

Figure 18. Experimental setup for the through-wall scene.

First of all, we directly utilized the CC-TDBP algorithm proposed in Section 2.1 to focus the echo data regardless of the wall effect. Background subtraction was applied to filter the reflections from the wall and the direct coupling among the antennas. It can be observed that the two targets in 2D imaging (Figure 19a) can be easily reconstructed and distinguished. The locations of the targets are marked by the red circle in Figure 19a, which verifies the imaging and detection performance of the designed MIMO radar system after the EM wave is attenuated due to the wall. However, it is worth noting that the wall effects, such as refraction dispersion and change of the EM wave speed, will impact the aforementioned data processing, generating an extra time delay and shifting of the targets. The intersection of white lines in Figure 19a is the real positions of the targets. It is obvious that the targets appear more distant compared to the intersection points.

Then, the improved BP imaging algorithm mentioned in Section 3 was applied, which considers the existence of the wall. The experimental result was improved and the corrected 2D imaging result is shown in Figure 19b. It can be observed that the positions of the targets in the image basically coincides with the intersection positions of the white lines. The reconstructed locations of the two human targets were much closer to the actual positions, which demonstrates the effectiveness of the improved BP algorithm for more accurate location of human targets behind the wall.

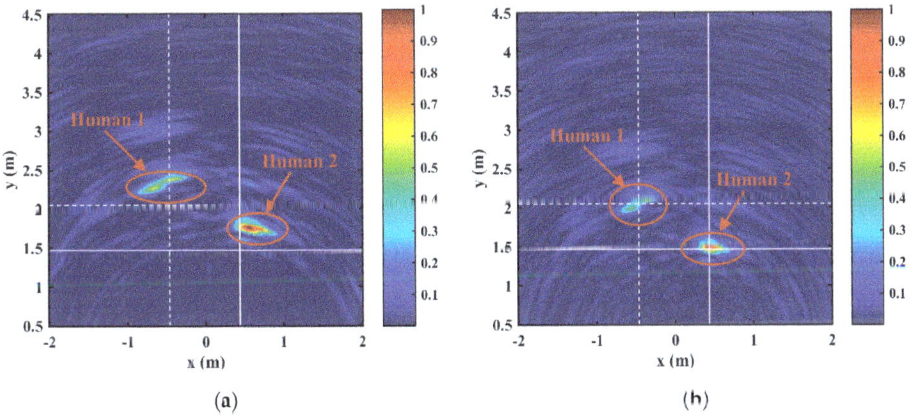

Figure 19. 2D through-wall imaging results for two humans behind a wall: (**a**) Imaging directly by the CC-TDBP algorithm with background removal; (**b**) Imaging by improved the CC-TDBP algorithm considering the presence of the wall.

5. Discussion

In this paper, a linear sparse MIMO radar system which can be used for through-wall detection was designed and analyzed. The array design methods can reduce the number of antennas while maintaining the azimuth resolution. The traditional Vivaldi antenna was improved by antenna miniaturization technology, so that the characteristics of high gain, ultra-wideband, and low frequency were realized. This miniaturization approach may also be utilized to improve the bandwidth characteristics of other similar printed antennas. Concerning the imaging, the TDBP imaging algorithm based on cross correlation was presented in detail. The experimental results in Figures 15 and 16 show that the proposed CC-TDBP algorithm can suppress artifacts better than the traditional TDBP algorithm. In addition, we considered the effect of the wall on the refraction of EM waves and the changes of velocity in the through-wall detection experiment (Section 4.2). Based on the principle of Snell's law and ray tracing, a new through-wall imaging model is established in Figure 12. The improved BP algorithm with cross-correlation compensates for the wall's influence and effectively focuses the targets behind the wall. Figure 19 shows that the targets' locations are closer to the real positions measured by meter stick. Finally, the significance and characteristics of this research are discussed in the following four points.

(1). The direct significance of the MIMO radar system studied in this paper is to improve traditional synthetic aperture methods. The designed radar system with optimized UWB Vivaldi antennas has the characteristics of fast data acquisition speed, required azimuth resolution, low cost, and can be applied to through-wall detection. The Vivaldi antenna has been designed, simulated, manufactured, and successfully used in through-wall imaging, which is one of the contributions to the through-wall radar technology. The period data (64 channels data) acquisition time of the MIMO radar system is approximately 35 s. However, this relatively long sampling time is due to the specific VNA that operated as the transceiver. It is an old model (Agilent N9925a) which is not designed for fast acquisition. In fact, the MIMO radar does not have mechanical moving parts, so it can acquire data much faster than synthetic aperture radar (SAR) based on the movement of the transceiver along a rail [17]. In addition, compared with the current research [9,37] which is limited to simulation and validation, we developed the actual radar system to carry out experimental verification, which is more conducive to illustrate the effectiveness of the proposed system design and imaging algorithm.

(2). Generally, the dielectric constant of the wall is larger than that of the air. The transmitted EM waves have to pass the wall twice to reach the radar receivers, which further reduces the energy of the received target signal. In terms of amplitude attenuation, the first reflection from the front wall is the strongest and the higher-order reflection can be neglected. The signal information of the targets hidden behind walls, such as back walls and human targets, are mainly conveyed by the first transmission. It is worth noting that the reflection of the human body is relatively low compared with wall reflection and the direct coupling between the antennas. Moreover, the attenuation of the wall makes the collected human target signal weaker, which certainly increases the difficulty of through-wall imaging [38]. Therefore, in the third experiment, background removal is used to improve the imaging performance.

(3). Based on the VNA platform, the system is more flexible. For example, we can properly change the operating frequency range and frequency points of the system according to the specific application scenario. Thus, the range resolution and maximum ambiguity range can be improved. In particular, when the wall is thicker, the low-frequency UWB range can be chosen to improve the penetration of EM waves.

(4). As the detection distance increases, the energy received by the radar system will inevitably be attenuated. As shown in Figures 15 and 19, when multiple targets are distributed at different distances in the imaging scene, the relatively weak energy of distant targets may be concealed in clutter. Therefore, we still need to study an effective way to solve this problem, such as adding amplifiers or shielding devices to the radar system to improve energy in the next step.

In future research, the application of MIMO through-wall radar to extract vital signs (such as breathing, heartbeat, arm swing) or human micro-Doppler [39,40] will be explored. In addition, compressive sensing technology can also be applied to through-wall detection of sparse arrays to reduce the sampling time and improve the resolution [41]. Moreover, it is worth mentioning that the existence of multiple walls, the detection of moving targets, real-time positioning imaging, and even three-dimensional imaging will also be challenging research tasks in our future work.

6. Conclusions

This paper develops a UWB MIMO radar system for through-wall imaging by the cross-correlation-based TDBP algorithm. Low-frequency UWB miniaturized Vivaldi antennas were designed and realized to ensure good system performance of wall penetration. The designed antenna, which is one of this paper's contributions, has acceptable through-wall imaging performance and is small in size, low in cost, and easy to manufacture. The system's working frequency is from 0.4 GHz to 2.6 GHz, conferring a range resolution of 6.8 cm. The aperture length of the designed MIMO array is about 1.1 m, resulting in an angle resolution of 5.8 rad. It is worth noting that the wall effects, such as refraction dispersion and change of the EM wave speed, are not ignored for the data processing. The through-wall imaging model is proposed and human targets were reconstructed more accurately in through-wall experiments. The results of the experiments demonstrate that, both in range and azimuth direction, the proposed imaging methods can effectively suppress artifacts and focus the different targets, and the designed MIMO radar system can detect and localize human targets behind a wall.

Author Contributions: Writing—original draft preparation, Z.H.; writing—review and editing, Z.H. and Z.Z.; methodology, Z.H., W.F. and K.W.; hardware design, Z.H., X.K. and W.F.; software, Z.H., W.F. and K.W.; validation, Z.H., J.Z. and Q.L.; formal analysis, Z.H., J.Z. and X.K.; data curation, Z.H., W.F. and K.W.; supervision, Q.L., Z.Z.; project administration, Z.Z.; funding acquisition, Z.Z.

Funding: This research was supported by the Natural Science Foundation of China (41174097, 41574097) and the Provincial School Construction Project of Jilin Province (SXGJSF 2017-5).

Acknowledgments: Zhipeng Hu is grateful to Tohoku University and Motoyuki Sato for a three-month visiting to Center for Northeast Asian Studies (CNEAS).

Conflicts of Interest: The authors declare no conflict of interest.

References

1. Amin, M.G. *Through-the-Wall Radar Imaging*; CRC Press: London, UK, 2011.
2. Baranoski, E.J. Through-wall imaging: Historical perspective and future directions. *J. Frankl. Inst.* **2008**, *345*, 556–569. [CrossRef]
3. Lan, F.; Kong, L.; Yang, X.; Jia, Y.; Ke, X. Life-sign detection of through-wall-radar based on fourth-order cumulant. In Proceedings of the 2013 IEEE Radar Conference, Ottawa, ON, Canada, 29 April–3 May 2013.
4. Amin, M.G. Radar, signal, and image processing techniques for through the wall imaging. In Proceedings of the Digital Wireless Communications VII and Space Communication Technologies, Orlando, FL, USA, 2 June 2005.
5. Nikolic, M.M.; Nehorai, A.; Djordjevic, A.R. Estimating moving targets behind reinforced walls using radar. *IEEE Trans. Antennas Propag.* **2009**, *57*, 3530–3538. [CrossRef]
6. Yılmaz, B.; Özdemir, C. Design and prototype of radar sensor with Vivaldi linear array for through-wall radar imaging: An experimental study. *J. Appl. Remote Sens.* **2016**, *10*, 046012. [CrossRef]
7. Ralston, T.S.; Charvat, G.L.; Peabody, J.E. Real-time through-wall imaging using an ultrawideband multiple-input multiple-output (MIMO) phased array radar system. In Proceedings of the 2010 IEEE international symposium on phased array systems and technology, Waltham, MA, USA, 12–15 October 2010.
8. Guo, S.; Cui, G.; Kong, L.; Song, Y.; Yang, X. Multipath Analysis and Exploitation for MIMO Through-the-Wall Imaging Radar. *IEEE J. Sel. Top. Appl. Earth Observ. Remote Sens.* **2018**, *11*, 3721–3731. [CrossRef]
9. Wang, M.; Cui, G.; Yi, W.; Kong, L.; Yang, X.; Yuan, T. Time-division MIMO through the-wall radar imaging behind multiple walls. In Proceedings of the 2015 IEEE Radar Conference, Arlington, VA, USA, 10–15 May 2015.
10. Dehmollaian, M.; Thiel, M.; Sarabandi, K. Through-the-wall imaging using differential SAR. *IEEE Trans. Geosci. Remote Sens.* **2009**, *47*, 1289–1296. [CrossRef]

11. Laviada, J.; Arboleya, A.; López-Gayarre, F.; Las-Heras, F. Broadband synthetic aperture scanning system for three-dimensional through-the-wall inspection. *IEEE Geosci. Remote Sens. Lett.* **2015**, *13*, 97–101. [CrossRef]

12. Yang, J.; Thompson, J.; Huang, X.; Jin, T.; Zhou, Z. Random-frequency SAR imaging based on compressed sensing. *IEEE Trans. Geosci. Remote Sens.* **2013**, *51*, 983–994. [CrossRef]

13. Jin, T.; Chen, B.; Zhou, Z. Image-domain estimation of wall parameters for autofocusing of through-the-wall SAR imagery. *IEEE Trans. Geosci. Remote Sens.* **2013**, *51*, 1836–1843. [CrossRef]

14. Bliss, D.W.; Forsythe, K.W. Multiple-input multiple-output (MIMO) radar and imaging: Degrees of freedom and resolution. In Proceedings of the Thrity-Seventh Asilomar Conference on Signals, Systems & Computers, Pacific Grove, CA, USA, 9–12 November 2003.

15. Fishler, E.; Haimovich, A.; Blum, R.; Chizhik, D.; Cimini, L.; Valenzuela, R. MIMO radar: An idea whose time has come. In Proceedings of the 2004 IEEE radar conference, Philadelphia, PA, USA, 29 April 2004.

16. Narayanan, R.M.; Gebhardt, E.T.; Broderick, S.P. Through-Wall Single and Multiple Target Imaging Using MIMO Radar. *Electronics* **2017**, *6*, 70. [CrossRef]

17. Pieraccini, M.; Miccinesi, L. An Interferometric MIMO Radar for Bridge Monitoring. *IEEE Geosci. Remote Sens. Lett.* **2019**, 1–5. [CrossRef]

18. Zhuge, X.; Yarovoy, A.G. Sparse multiple-input multiple-output arrays for high-resolution near-field ultra-wideband imaging. *IET Microw. Antennas Propag.* **2011**, *5*, 1552–1562. [CrossRef]

19. Zhuge, X. Short-Range Ultra-Wideband Imaging with Multiple-Input Multiple-Output Arrays. Ph.D. Thesis, Delft University Technology (TUDelft), Delft, The Netherlands, 2010.

20. Feng, W.; Zou, L.; Sato, M. 2D imaging by sparse array radar system. *IEICE Tech. Rep.* **2016**, *116*, 65–70.

21. Schwartz, J.L.; Steinberg, B.D. Ultrasparse, ultrawideband arrays. *IEEE Trans. Ultrason. Ferroelectr. Freq. Control* **1998**, *45*, 376–393. [CrossRef] [PubMed]

22. Maaref, N.; Millot, P. Array-based UWB FMCW through-the-wall radar. In Proceedings of the 2012 IEEE International Symposium on Antennas and Propagation, Chicago, IL, USA, 8–14 July 2012.

23. Fioranelli, F.; Salous, S.; Ndip, I.; Raimundo, X. Through-the-wall detection with gated FMCW signals using optimized patch-like and Vivaldi antennas. *IEEE Trans. Antennas Propag.* **2015**, *63*, 1106–1117. [CrossRef]

24. Gibson, P.J. The vivaldi aerial. In Proceedings of the 1979 9th European Microwave Conference, Brighton, UK, 17–20 September 1979.

25. Fei, P.; Jiao, Y.C.; Hu, W.; Zhang, F.S. A miniaturized antipodal Vivaldi antenna with improved radiation characteristics. *IEEE Antennas Wirel. Propag. Lett.* **2011**, *10*, 127–130.

26. Rizk, J.B.; Rebeiz, G.M. Millimeter-wave Fermi tapered slot antennas on micromachined silicon substrates. *IEEE Trans. Antennas Propag.* **2002**, *50*, 379–383. [CrossRef]

27. Feng, W.; Yi, L.; Sato, M. Near range radar imaging based on block sparsity and cross-correlation fusion algorithm. *IEEE J. Sel. Top. Appl. Earth Observ. Remote Sens.* **2018**, *11*, 2079–2089. [CrossRef]

28. Zhou, L.; Huang, C.; Su, Y. A fast back-projection algorithm based on cross correlation for GPR imaging. *IEEE Geosci. Remote Sens. Lett.* **2012**, *9*, 228–232. [CrossRef]

29. Gumbmann, F.; Tran, P.; Schmidt, L.P. Sparse linear array design for a short range imaging radar. In Proceedings of the 2009 European Radar Conference, Rome, Italy, 30 September–2 October 2009.

30. Ge, T.; Zhao, L.; Cai, Y.; Zhou, J. A grating lobes suppression technique for near-field sparse linear MIMO array imaging. In Proceedings of the 2016 CIE International Conference on Radar, Guangzhou, China, 10–13 October 2016.

31. Jin, L.; Ouyang, S.; Zhou, L. Array design and imaging method for ultra-wideband multiple-input multiple-output through-the-wall radar. *J. Electron. Inf. Technol.* **2012**, *34*, 1574–1580. (In Chinese) [CrossRef]

32. Fishler, E.; Haimovich, A.; Blum, R.; Cimini, L.; Chizhik, D.; Valenzuela, R. Spatial diversity in radars-models and detection performance. *IEEE Trans. Signal Process.* **2006**, *54*, 823–838. [CrossRef]

33. Bellettini, A.; Pinto, M.A. Theoretical accuracy of synthetic aperture sonar micronavigation using a displaced phase-center antenna. *IEEE J. Ocean. Eng.* **2002**, *27*, 780–789. [CrossRef]

34. Teni, G.; Zhang, N.; Qiu, J.; Zhang, P. Research on a novel miniaturized antipodal Vivaldi antenna with improved radiation. *IEEE Antennas Wirel. Propag. Lett.* **2013**, *12*, 417–420. [CrossRef]

35. Guo, Z.; Yang, S.; Shi, Z.; Chen, Y. A miniaturized wideband dual-polarized linear array with balanced antipodal Vivaldi antenna. In Proceedings of the 2016 IEEE MTT-S International Microwave Workshop Series on Advanced Materials and Processes for RF and THz Applications (IMWS-AMP), Chengdu, China, 20–22 July 2016.

36. Yang, Y.; Wang, Y.; Fathy, A.E. Design of compact Vivaldi antenna arrays for UWB see through wall applications. *Prog. Electromagn. Res.* **2008**, *82*, 401–418. [CrossRef]

37. Wang, K.; Zeng, Z.; Sun, J. Through-Wall Detection of the Moving Paths and Vital Signs of Human Beings. *IEEE Geosci. Remote Sens. Lett.* **2018**, *16*, 717–721. [CrossRef]

38. Song, Y.; Hu, J.; Chu, N.; Jin, T.; Zhang, J.; Zhou, Z. Building Layout Reconstruction in Concealed Human Target Sensing via UWB MIMO Through-Wall Imaging Radar. *IEEE Geosci. Remote Sens. Lett.* **2018**, *15*, 1199–1203.

39. Qi, F.; Liang, F.; Lv, H.; Li, C.; Chen, F.; Wang, J. Detection and Classification of Finer-Grained Human Activities Based on Stepped-Frequency Continuous-Wave Through-Wall Radar. *Sensors* **2016**, *16*, 885. [CrossRef] [PubMed]

40. Gennarelli, G.; Ludeno, G.; Soldovieri, F. Real-Time Through-Wall Situation Awareness Using a Microwave Doppler Radar Sensor. *Remote Sens.* **2016**, *8*, 621. [CrossRef]

41. Ma, Y.; Hong, H.; Zhu, X. Interaction Multipath in Through-the-Wall Radar Imaging Based on Compressive Sensing. *Sensors* **2018**, *18*, 549.

 *remote sensing*

Article

On the Capabilities of the Italian Airborne FMCW AXIS InSAR System

Carmen Esposito [1], Antonio Natale [1,*], Gianfranco Palmese [2], Paolo Berardino [1], Riccardo Lanari [1] and Stefano Perna [1,3]

[1] Institute for Remote Sensing of Environment (IREA), National Research Council (CNR), 80124 Napoli, Italy; esposito.c@irea.cnr.it (C.E.); berardino.p@irea.cnr.it (P.B.); lanari.r@irea.cnr.it (R.L.); perna@uniparthenope.it (S.P.)
[2] Elettra Microwave, 80143 Napoli, Italy; g.palmese@elettramicrowave.it
[3] Department of Engineering (DI), Università degli Studi di Napoli "Parthenope", 80143 Napoli, Italy
* Correspondence: natale.a@irea.cnr.it; Tel.: +39-081-762-0632

Received: 31 December 2019; Accepted: 4 February 2020; Published: 6 February 2020

Abstract: Airborne Synthetic Aperture Radar (SAR) systems are gaining increasing interest within the remote sensing community due to their operational flexibility and observation capabilities. Among these systems, those exploiting the Frequency-Modulated Continuous-Wave (FMCW) technology are compact, lightweight, and comparatively low cost. For these reasons, they are becoming very attractive, since they can be easily mounted onboard ever-smaller and highly flexible aerial platforms, like helicopters or unmanned aerial vehicles (UAVs). In this work, we present the imaging and topographic capabilities of a novel Italian airborne SAR system developed in the frame of cooperation between a public research institute (IREA-CNR) and a private company (Elettra Microwave S.r.l.). The system, which is named AXIS (standing for Airborne X-band Interferometric SAR), is based on FMCW technology and is equipped with a single-pass interferometric layout. In the work we first provide a description of the AXIS system. Then, we describe the acquisition campaign carried out in April 2018, just after the system completion. Finally, we perform an analysis of the radar data acquired during the campaign, by presenting a quantitative assessment of the quality of the SLC (Single Look Complex) SAR images and the interferometric products achievable through the system. The overall analysis aims at providing first reference values for future research and operational activities that will be conducted with this sensor.

Keywords: Synthetic Aperture Radar (SAR); Airborne SAR; SAR Interferometry; Digital Elevation Model (DEM); Frequency-Modulated Continuous-Wave (FMCW)

---

1. Introduction

Synthetic Aperture Radar (SAR) systems are microwave remote sensors that are mounted on board moving platforms in order to obtain high spatial resolution in the along-track direction by emulating the acquisition mechanism of large-aperture antennas [1,2]. Very common platforms used to mount SAR systems are satellites [1–6] (spaceborne systems), airplanes [7–10], helicopters [11], and, more recently, drones [12,13] (aerial systems). Due to their peculiarities, spaceborne and aerial SAR systems are, to some extent, complementary.

Spaceborne systems guarantee very wide spatial coverage. However, they are forced to follow polar orbits, thus flying practically only along the South–North (or North–South) direction. This poses some limitations to the full exploitation of those techniques, such as Differential SAR Interferometry (DInSAR) or Along Track Interferometry (ATI), which allow us to measure only the Line of Sight (LoS) component of the remotely sensed phenomenon. Moreover, with the currently operative spaceborne SAR constellations [3,4], the revisiting time, that is, the time interval elapsing between subsequent observations of the same area, is on the order of several days. This makes it impossible on the one hand to illuminate the area of interest in a timely way in case of emergencies, and on the other hand to monitor the evolution of the remotely sensed phenomenon through daily or hourly observations.

On the contrary, aerial systems guarantee narrow spatial coverage. However, they can fly in any direction and, at least in principle, whenever required. Accordingly, they allow us to reach the area to observe in a timely way, and to reduce the revisiting time to a few minutes. Furthermore, aerial systems can use antennas much smaller than those installed on spaceborne platforms, due to the significantly reduced distance from the observed targets. This ensures a geometrical resolution in the along-track direction higher than that achievable with the spaceborne systems [1,2]. In addition, when high-frequency bands (like Ka, Ku, X, and C) are employed, aerial platforms allow us to easily obtain effective single-pass InSAR configurations [8–10,12,14–19]. Indeed, considering the usual flight altitudes of aerial platforms, for wavelengths on the order of a few centimeters (or less), the interferometric height of ambiguity [1,2] can be kept sufficiently low with baselines [1,2] as small as required by the geometrical constraints imposed by small aerial platforms. In this regard, it is recalled that aerial single-pass InSAR configurations are particularly attractive for two reasons. First, because (like any single-pass configuration) they allow for circumventing temporal decorrelation effects [1,2]. Second, because they allow us to strongly mitigate in the interferometric products the influence of the so-called residual errors which are typical of aerial SAR focused images due to the unavoidable inaccuracies of the navigation data used during the image formation procedure [20].

The complementary peculiarities of spaceborne and aerial SAR systems have led in the last years to two contrasting trends, which have somehow driven the technological development of these systems.

The first trend answers to the requirement of illuminating larger and larger areas. To do this, spaceborne SAR systems are appropriate. In this case, the technological challenge to face consists of the implementation of advanced acquisition modes, such as ScanSAR [1,2,21,22], TOPS [2,5,22,23], and/or advanced optimization strategies, such as the digital beam-forming on receive technique [24], which are aimed at widening the across-track (XT) coverage achievable with the more conventional Stripmap mode [1,2]. This trend, of course, enables the growth of novel methods of data reduction and analysis by artificial intelligence (AI), due to the sharply increasing spaceborne SAR data volume [25–28].

The second trend instead follows the need for guaranteeing fast and flexible monitoring, possibly at high resolution, of confined areas. For this purpose, aerial systems are appropriate. In this case, the technological challenge to face consists of the reduction of the size, weight, and realization costs of the developed SAR system. In this frame, beside the conventional pulse radar systems, Frequency-Modulated Continuous-Wave (FMCW) [29,30] is emerging as a very attractive solution. Indeed, unlike the pulse radar systems, which require high peak transmission power, the FMCW systems operate with constant low transmission power. In addition, the sampling frequency of the Analog-to-Digital Converter (ADC) of the FMCW SAR systems can be significantly smaller than the bandwidth of the transmitted signal. On the other hand, the operating principle of the FMCW SAR limits the maximum detectable sensor-to-target distance to a few kilometers, which is safely acceptable for the acquisition geometry of several aerial platforms, especially for the small-sized ones, which typically fly at very low altitudes. Summing up, the FMCW SAR systems are particularly tailored to small aerial platforms, since their architecture complexity, which is lower than that of the pulse SAR systems, involves a reduction of size, weight, and realization costs.

In this frame, the Institute for the Electromagnetic Sensing of the Environment (IREA) of the National Italian Research Council (CNR) has recently signed an agreement with "Elettra Microwave", which is a small Italian company, for the scientific use of a novel single-pass interferometric airborne FMCW SAR prototype realized by the company. The system is named AXIS, which stands for Airborne X-band Interferometric System. Like any conventional FMCW radar, it operates in a bistatic configuration. Therefore, to obtain a single-pass interferometric layout, it mounts three radar antennas: one transmitting (Tx) and two receiving (Rx).

In this work, we present a first assessment of the imaging and topographic mapping capabilities of the AXIS system. To do this, we show the results relevant to the acquisition campaign carried out over the Salerno area, South of Italy, in 2018 just after system completion. In particular, during the campaign, a number of Corner Reflectors (CRs) were deployed within the area illuminated by the radar, and very accurate measurement of their positions through the Differential Global Positioning System (D-GPS) technique was carried out to provide a set of sound reference ground points. This allowed a first assessment of the quality of the focused SAR images and the Interferometric SAR (InSAR) products achieved with the AXIS system. More specifically, in correspondence with this set of reference points, the geometric resolution and the planar positioning accuracy of the focused AXIS images were measured. Then, a comparison between the DGPS measurements of the CRs' positions and the Digital Elevation Model (DEM) generated with the single-pass InSAR AXIS data was carried out.

The presented analysis aims at providing first reference values for future research and operational activities that will be conducted with this sensor.

The work is organized as follows. In Section 2 we provide a brief description of the system. The acquisition campaign, the processing chain applied to the SAR data, and the achieved results are described in Section 3. The concluding remarks are reported in Section 4.

2. System Description

The AXIS system basically consists of a radar module (accommodated in a rack), which embeds an accurate navigation unit, and three different radar antennas. The system is currently mounted on board a Cessna 172 aircraft, whose main parameters are collected in Table 1.

Table 1. Airplane parameters.

Model	Cessna 172
Propulsion	1 Lycoming IO-360-L2A
Velocity	up to 228 km/h
Endurance	5 h

The description of the main modules of the system is provided in the following subsections.

2.1. Navigation Unit

In order to limit the effects of the residual errors which are typical of airborne SAR data [20], the AXIS system uses a very precise navigation unit, namely, the Applanix POS-AV510, which contains a Global Navigation Satellite System (GNSS) and an Inertial Measurement Unit (IMU), which is directly connected to the radar module and accommodated on the top of the rack. The system was originally mounted on board the InSAS4 airborne SAR [9]. As clarified in [9], the use of this navigation unit, coupled to proper postflight processing techniques, guarantees very precise flight parameter measurability, as reported in Table 2.

Table 2. Absolute Accuracy Specifications (RMS) [+] of the Inertial Measurement Unit (IMU) [*].

Position	0.05 m
Velocity	0.005 m/s
Roll and Pitch	0.005°
True Heading	0.008°

[+] Root Mean Square. [*] after post-processing integration with GNSS data.

2.2. Antennas

The AXIS system mounts three X-Band commercial off-the-shelf VV polarized microstrip antennas. They are very compact and light: each of them is mounted in a 24.6 × 12.6 × 1.5 cm radome with an overall weight of 420 g.

The antenna parameters were measured in the anechoic chamber of the Università degli Studi di Napoli "Parthenope". They are practically the same for the three antennas. As an example, we report in Figure 1 the elevation and azimuth cuts of the (normalized) radiation pattern relevant to one of the three antennas, namely that with serial number MAA-935985-V. In this case, the measured gain [31] is 24.7 dB; the measured half power beamwidth [31] is equal to 19.2° in elevation and 7.4° in azimuth. When considering the squared pattern (to account for the two-way signal path), these values decrease to 14° and 5.4°, respectively.

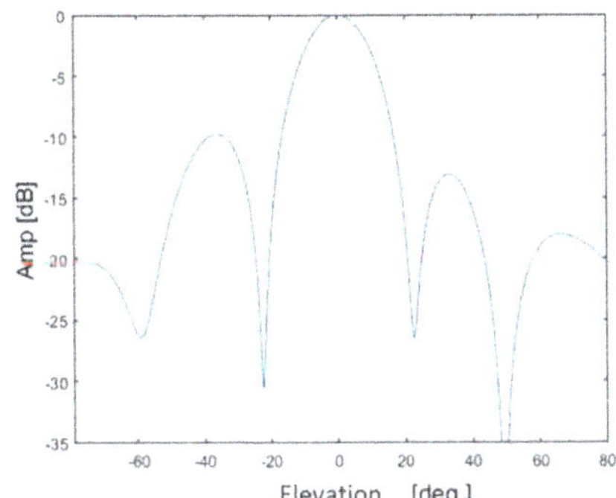
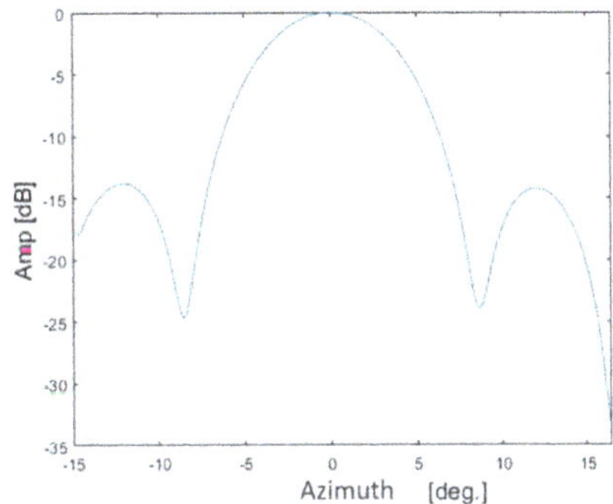

Figure 1. Elevation (**left**) and azimuth (**right**) normalized radiation patterns of one of the three radar antennas of the AXIS (Airborne X-band Interferometric Synthetic Aperture Radar (SAR)) system.

It is finally remarked that in the anechoic chamber was also carried out a very accurate measurement of the phase center position of each radar antenna, see [32].

2.3. Interferometric Layout

The AXIS system is equipped with a single-pass interferometric layout. As remarked above, like any conventional FMCW system, AXIS is a bistatic radar; that is, the Tx antenna is not also used to receive. Accordingly, three radar antennas, one Tx and two Rx, are necessary to obtain a single-pass interferometric configuration.

As specified above, the system is currently mounted on board a Cessna 172 aircraft. For this airplane, the three antennas are mounted on the strut of the right wing by means of customized camera mounts; see Figure 2. By doing so, the antennas exhibit a pointing angle in the elevation plane equal to 45 degrees; moreover, the interferometric baseline between the two Rx devices is equal to 1.75 m. For flight altitudes of 1 Km, 2 Km, and 3 Km, which are typical of comparatively small aircrafts, this baseline leads to the height of ambiguity values [1] reported in Figure 3.

The lever arms, namely, the distances between the antennas' phase centers and the reference center of the navigation unit, were measured using the theodolite technique before the airborne mission [9,33].

Some considerations on the strategy adopted to obtain the AXIS interferometric layout are now in order.

First of all, we underline that one of the reasons why we chose to mount the antennas on the wing strut is that this is one of the least flexible structures of the aircraft. This allows us to limit the problems related to the presence of small deformations unavoidably occurring on the aircraft body during flight. We are thus quite confident that the difference between the in-flight and measured lever arms is negligible; in any case, this difference is reasonably within the accuracy of the adopted theodolite measurement equipment.

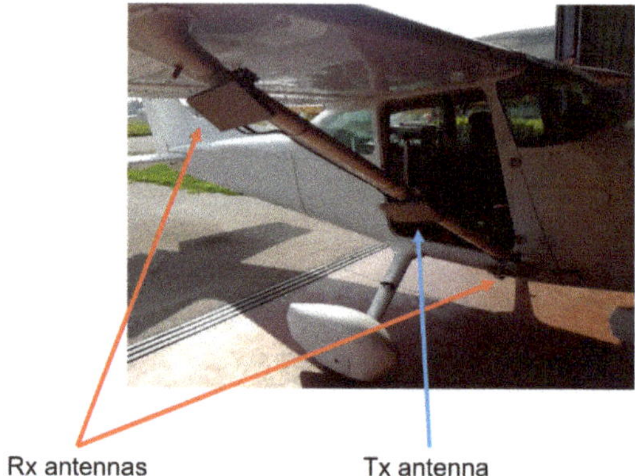

Figure 2. Interferometric layout of the AXIS system. The light blue arrow points to the transmitting radar antenna, whereas the red arrows point to the two receiving radar antennas.

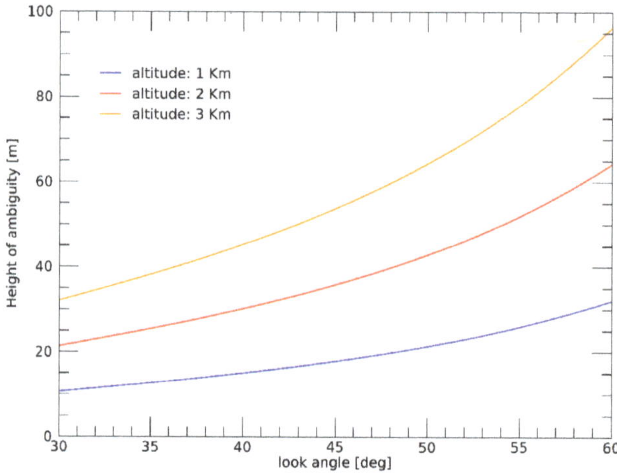

Figure 3. Height of ambiguity of the AXIS system versus the look angle for three different flight altitudes.

Second, we remark that the three camera mounts shown in Figure 2 are fixed separately. Accordingly, the solution adopted to install the three antennas could not ensure sufficiently accurate (that is, within the accuracy of the adopted theodolite measurement equipment) repeatability of the overall obtained antenna layout (that is, the relative positions of the antennas as well as their orientations in azimuth and elevation). For this reason, with the architectural solution shown in Figure 2 it is preferable to repeat the measurement of the lever arms for each acquisition campaign. This, of course, may represent a strong limitation on the flexible use of the radar, especially if we intend to mount it on board the aircraft only when required. Indeed, in an operative crisis scenario, in which fast data acquisition and processing could be requested, the lever arms measurement operation could dramatically delay the beginning of the flight mission and data processing (we recall that precise knowledge of the lever arms is necessary to ensure accurate airborne SAR data focusing [20]). To overcome this limitation, for future missions, a rigid mechanical framework embedding the three antennas (and tailored to any Cessna 172 model) was designed and built. In this way, at worst a constant (and reasonably small) bias between the overall antennas' frame and the IMU reference center will occur when re-installing the antennas before each mission. We are quite confident that this will make it unnecessary to measure the lever arms after each antenna installation. In any case, a deeper analysis of these kinds of issues is a matter of current investigation.

2.4. Radar

The AXIS radar exploits FMCW technology, which allows for operating in a high-resolution mode though exploiting the comparatively low ADC sampling frequency and low data rate. The main radar parameters are collected in Table 3. Note in particular that, following the notation largely used in the literature [29,30], in the table we have used some terms (Pulse repetition interval, Pulse duration, Pulse repetition frequency) borrowed from pulse radar jargon. The meaning of these terms for FMCW radar is clarified in Figure 4, which shows the temporal behavior of the frequency of the signal continuously transmitted by such kinds of radars. In this regard, note in particular that the pulse repetition frequency is equal to the inverse of the pulse repetition interval. In the specific case of the AXIS system, the recording data time is 605 µs, while the frequency sweep rate, accurately measured following the procedure in [34], is 3.336×10^{11} s^{-2}, which leads to a maximum range distance of about 5615 m recordable by the radar. Moreover, the sampling rate is 25 MHz, while the bandwidth of the transmitted signal is 200 MHz, leading to a slant range resolution of about 0.75 m [30].

Table 3. Radar parameters.

Radar technology	FMCW
Transmitted power	5 W
Carrier frequency	9.55 GHz
Bandwidth	200 MHz
Pulse repetition frequency	1200 Hz
Pulse repetition interval	833.33 µs
Pulse duration	600.184 µs
Recording data time	605.00 µs
Sampling rate	25 MHz
Frequency sweep rate	3.336×10^{11} s^{-2}
Range pixel spacing	0.74 m
Maximum recordable range	5615 m
Number of antennas	3
Polarization	VV

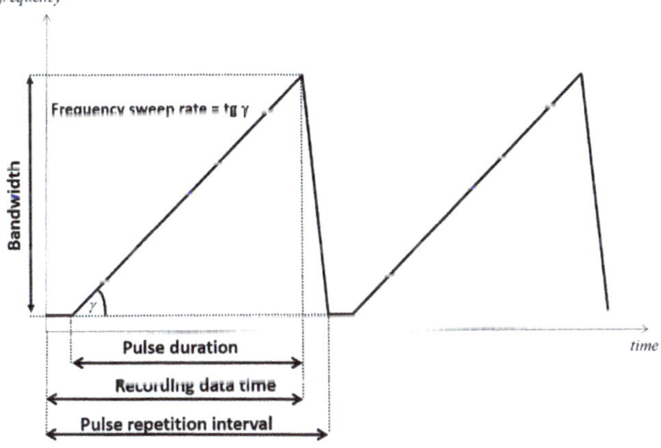

Figure 4. Operating principle of a FMCW SAR.

The Noise Equivalent Sigma Zero (NESZ) [29] of the system was computed according to the antenna gain measurements carried out in the laboratory (see previous section) and the radar components' specifications. The behavior of the NESZ versus the radar look angle is reported in Figure 5 for different flight altitudes.

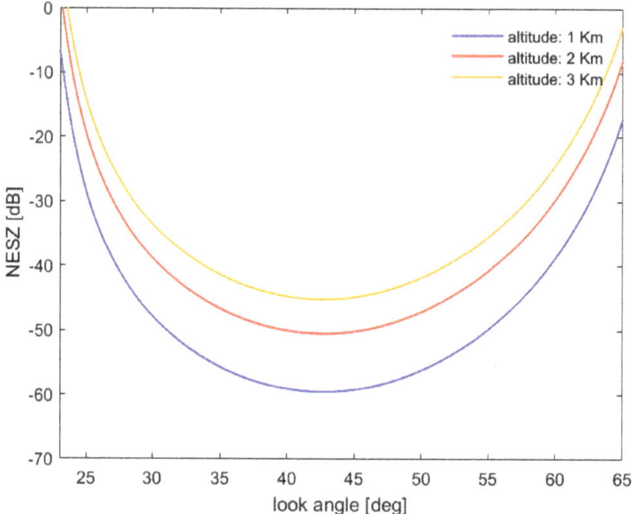

Figure 5. AXIS Noise Equivalent Sigma Zero (NESZ) as a function of the look angle for different flight altitudes.

A block diagram of the radar module is shown in Figure 6. As can be seen in the figure, it consists of three main blocks, namely, the Radar Digital Unit (RDU), the Radio Frequency Unit (RFU), which is connected to the three antennas, and the Power Supply Unit (PSU). In particular, the RDU is fully programmable and allows for setting the acquisition parameters, the generation timing, and the data handling. It also includes the ADC and the data storage unit. The RFU includes the frequency generation unit (which generates all the synchronization and radio frequency signals) and the chirp generator unit (which generates the low-frequency modulated chirp signal by means of digital direct synthesis technology). The RFU also includes the power amplifier, the Up/Down converter, and an antenna front-end. The PSU provides the power supply to the whole system by 24–28 V DC internal power. It is completely autonomous and does not affect the aircraft DC bus power. The navigation system is directly connected to the RDU by means of a specific interface; all the navigation data are synchronized with the radar pulses and embedded in the output data. The radar module has a weight of approximately 30 kg (comprising the PSU), and it is accommodated in a rack whose size is about 50 cm × 50 cm × 65 cm. Due to the comparatively low weight and compactness of the radar module, the AXIS system can be mounted on board small aircraft and helicopters. As specified above, the system is currently mounted on board a Cessna 172 aircraft. In particular, the rack that includes the radar module and the navigation system is installed in the aircraft cabin, in place of the two rear passenger seats.

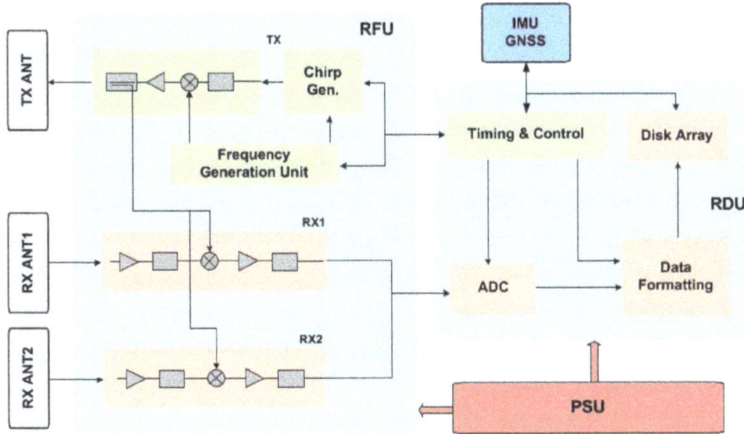

Figure 6. Block diagram of the AXIS system.

3. Experimental Results

In this section, we present the results relevant to the SAR data acquisition campaign carried out using the AXIS system over the Salerno area, Italy, in April 2018. More specifically, a flight campaign consisting of six overlapping flight circuits was scheduled, each of them containing two antiparallel linear tracks of about 20 km. In other words, we planned to collect SAR data from 12 flight tracks: 6 overlapping tracks from southeast (SE) to northwest (NW) and 6 overlapping tracks from NW to SE. In Figure 7, which shows an optical image of the test area, the actually flown circuits (obtained through the navigation data recorded during the overall campaign) are depicted with yellow lines.

Figure 7. The flight circuits (yellow lines) flown over the Salerno area, Italy, during the AXIS acquisition campaign.

Besides the flight campaign, we also performed a ground campaign aimed first of all at measuring the antennas' lever arms, in order to provide the information necessary to accurately process the radar data [35,36]. As specified above, such measurements were carried out very precisely through a Total Station Theodolite. The ground campaign was also aimed at providing a number of sound ground control points to assess the quality of the obtained SAR images and interferometric products. To do this, 10 CRs (5 for the NW–SE and 5 for the SE–NW flight tracks) were deployed over the area illuminated by the radar, and their positions were accurately measured by means of D-GPS surveys.

The main SAR acquisition parameters are summarized in Table 4. We recall that the minimum recordable range for an FMCW system is 0 m. In our case, we focused only a range portion of the overall acquired data by setting a near (slant) range equal to about 2478 m, corresponding to a (mean) look angle of about 20 degrees.

Table 4. SAR acquisition parameters.

Flight altitude	2500 m
Mean platform velocity	48 m/s
Azimuth pixel spacing *	0.04 m

* Raw data.

In Figure 8, a block diagram of the adopted InSAR processing chain is depicted. In particular, the range compression of FMCW SAR data simply requires a Fourier transform of each range line [30]. The azimuth compression step was carried out through a time-domain Back Projection (BP) strategy [35,36] by exploiting the information provided by the measured antennas' phase centers and lever arms, the navigation data and an external DEM, namely the SRTM one [37], of the observed area. The adopted processing strategy allowed us to avoid application of the approximations [38] necessary to implement frequency-domain focusing approaches with integrated motion compensation [39,40]. This, of course, involves an increase of the computational burden, but this can be managed by means of parallel computing strategies that, for the time-domain approaches, are very easy to implement. Moreover, like all the time-domain focusing algorithms based on BP approaches, our processing strategy allowed us to focus all the SAR images in a common output grid, thus avoiding the need to apply the co-registration step [1] to generate the SAR interferograms. Except for this latter processing step, standard InSAR processing [1] was applied for the generation of the interferometric products. Hereafter, we focus our attention on a single-pass interferometric dataset related to a flight track flown from SE to NW. More specifically, we first analyze the obtained amplitude SAR images and then the interferometric products.

Figure 9 shows the amplitude of the multi-look complex (MLC) SAR image relevant to the whole track. The image is focused in an output grid coincident with the radar one (that is, slant range and azimuth). Note that in the right vertical axis of the figure is specified the (mean) look angle corresponding to the range coordinate reported in the left vertical axis. We remark that in the figure, a 10 range × 10 azimuth pixel averaging window was applied for visualization purposes, obtaining 15 m × 16 m pixel spacing. The details of the main processing parameters are reported in Table 5. From the figure we note amplitude decay for small values (approximately less than 25 degrees) and high values (approximately greater than 60 degrees) of the (mean) look angle. This is in agreement with the NESZ curves of Figure 5 (we recall that for the considered flight altitude, the region of our interest in Figure 5 lies between the red and yellow curves).

Figure 9b instead reports a multi-look amplitude SAR image of a small area around Salerno's airport (see the light blue box in Figure 9a), which was processed at a higher resolution (see again Table 5). Note, in particular, that a 2 range × 10 azimuth pixel averaging window was applied in the figure for visualization purposes, obtaining 1.5 m × 1.6 m pixel spacing. All the five deployed CRs relevant to the SE–NW track are present in this area; they are highlighted in the figure with red circles.

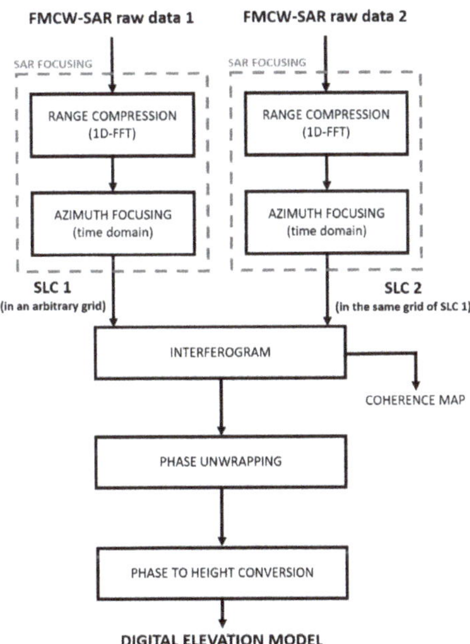

Figure 8. Block diagram of the adopted airborne InSAR processing chain.

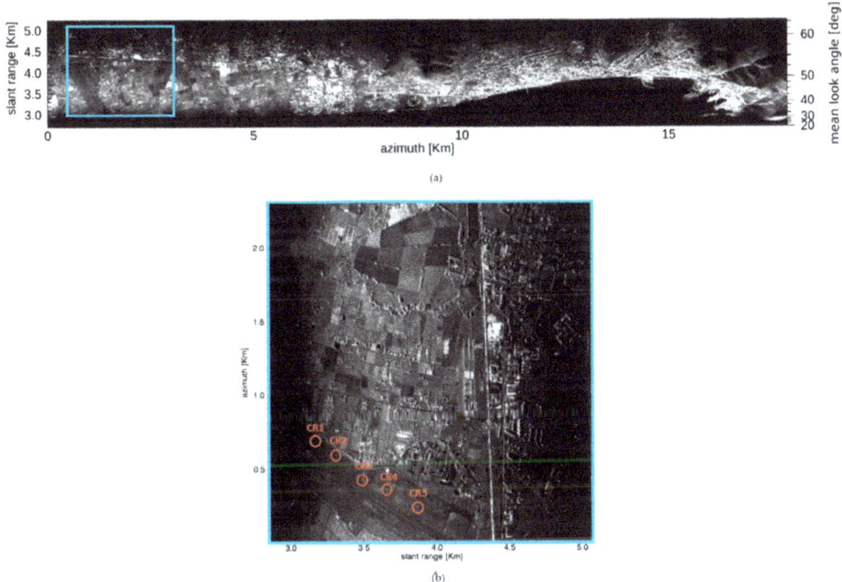

Figure 9. (a) Multi-look amplitude SAR images relevant to the acquired data. A 10 range × 10 azimuth pixel averaging window was applied, obtaining 15 m × 16 m pixel spacing. (b) Multi-look amplitude image of a patch of the entire acquired strip. A 2 range × 10 azimuth pixel averaging window was applied, obtaining 1.5 m × 1.6 m pixel spacing. Red circles indicate the Corner Reflectors' (CRs') positions.

Table 5. SAR data processing parameters relevant to the results collected in Figure 9a,b.

Parameters	Figure 9a	Figure 9b
Azimuth sampling (output grid)	1.6 m	0.16 m
Azimuth resolution	1.75 m	0.33 m
Azimuth resolution (MLC)	16 m	1.6 m
Range sampling (output grid)	1.5 m	0.75 m
Range resolution	1.5 m	0.75 m
Range resolution (MLC)	15 m	1.5 m

Processing parameters relevant to both images are listed in Table 5. For both images, the output grid coincides with the radar one (range–azimuth). Starting from the single-pass data pair focused with the processing parameters of Figure 9a, we obtained the wrapped interferogram and the corresponding coherence map shown in Figures 10a and 11a, respectively. Note that, as in Figure 9a, a 10 range × 10 azimuth pixel averaging window was applied to generate these two maps. Figures 10b and 11b instead report the interferogram and the corresponding coherence map obtained starting from the single-pass data pair focused with the processing parameters of Figure 9b. As in Figure 9b, a 2 range × 10 azimuth pixel averaging window was applied to generate these two maps. It is stressed that to obtain the interferometric products shown in Figures 10 and 11, besides the averaging window described above, we did not apply any additional filter aimed at limiting the noise effects. It is finally noted that in all the interferograms shown, we removed the topographic component provided by the external SRTM DEM used during the focusing step.

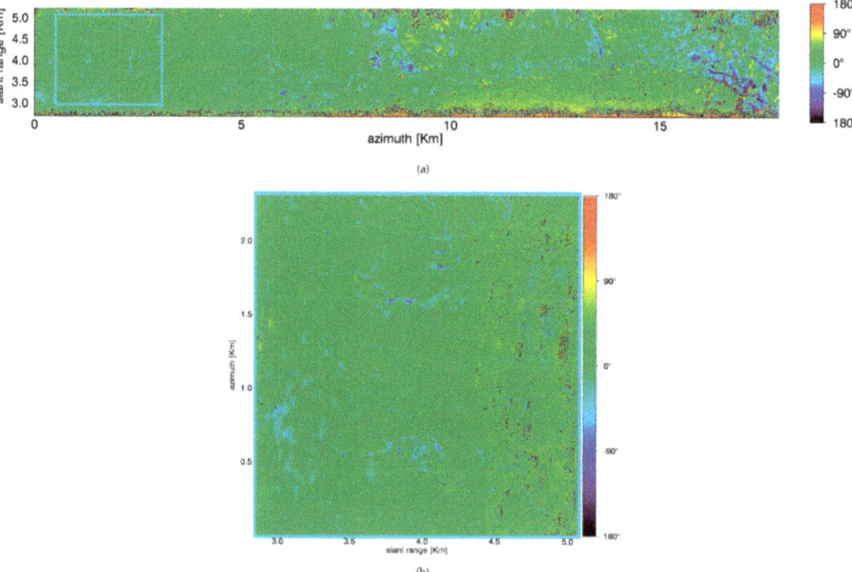

Figure 10. (**a**) Multi-look interferogram obtained from the single-pass data pair focused with the processing parameters of Figure 9a. A 10 range × 10 azimuth pixel averaging window was applied, obtaining 15 m × 16 m pixel spacing. (**b**) Multi-look interferogram obtained from the single-pass data pair focused with the processing parameters of Figure 9b. A 2 range × 10 azimuth pixel averaging window was applied, obtaining 1.5 m × 1.6 m pixel spacing.

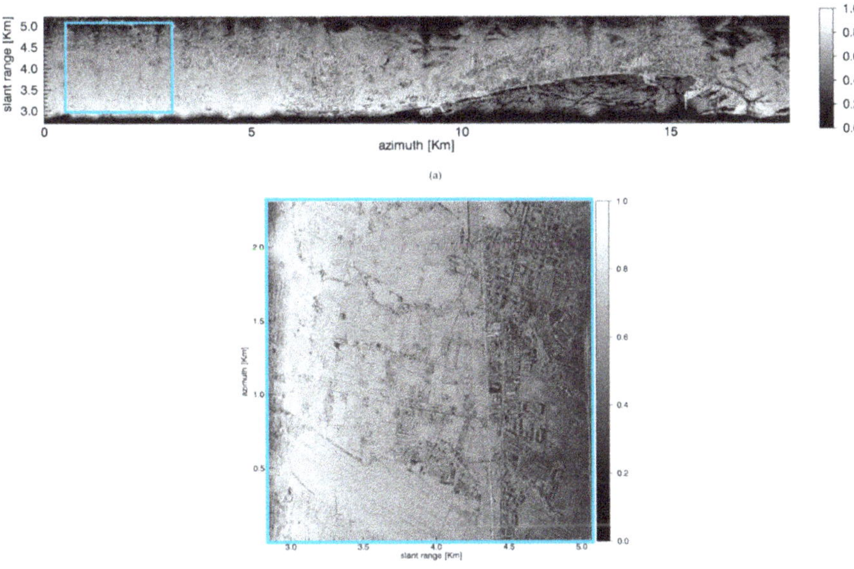

Figure 11. (a) Coherence map relevant to the interferogram of Figure 10a. (b) Coherence map relevant to the interferogram of Figure 10b.

Starting from the wrapped interferograms of Figure 10, we applied the phase unwrapping procedure from [41,42]. Then, we estimated the resulting unknown phase offset present in the unwrapped interferograms by applying the Phase-Based Estimate (PBE) procedure detailed in [33,43,44] and exploiting the D-GPS measurements relevant to the CRs. Thereafter, we carried out the phase-to-height conversion [1] to generate the InSAR DEM. The result relevant to the low-resolution interferogram of Figure 10a is displayed in Figure 12 on a geographic grid and superimposed upon an optical image of the overall test area.

Figure 12. InSAR DEM obtained starting from the low-resolution interferogram considered in Figure 10a. The DEM was geocoded and superimposed over a Google Earth image.

A quantitative assessment of the presented SAR products was performed by exploiting the five CRs shown in Figure 9b, along with the in situ D-GPS measurements relevant to their positions. More specifically, we carried out three different experiments by exploiting the high-resolution SAR image and interferogram generated with the processing parameters considered in Figures 9b and 10b.

In the first experiment, we measured the geometric resolution of the Single Look Complex (SLC) image in correspondence with the five CRs. The achieved results are listed in Table 6. We recall that the expected values are 0.33 m in azimuth and 0.75 m in range (see Table 5). In particular, we measured a mean azimuth resolution of 0.36 m with a standard deviation of 0.008 m, and a mean range resolution of 0.74 m with a standard deviation of 0.04 m. Note that no filtering (such as Hamming [1]) aimed at reducing the side lobe level of the point spread function (PSF) was applied. It is evident from Table 6 that the measured resolutions are very close to the expected theoretical ones. Moreover, it can also be noted that in most cases (with the exception of CR3) the measured range resolution is finer than the expected theoretical one. This is due to the fact that we chose the output geometry of the focused image according to the mean antenna pointing direction along the azimuth direction. Accordingly, the output geometry (also named processing geometry in the literature [45]) may be different from the acquisition geometry dictated by the actual antenna pointing direction within the azimuth aperture from which a generic target is illuminated. When this happens, the two-dimensional PSF relevant to the target is rotated with respect to the output grid [45]. As a consequence, measuring the resolution of the two-dimensional PSF along the azimuth and range directions of the output grid leads to an apparent improvement of the lower resolution (in our case, the range one; see Table 5) and an apparent impairment of the higher resolution (in our case, the azimuth one; see Table 5).

Table 6. Measurements on CRs.

	Azimuth Resolution [m]	Range Resolution [m]	Azimuth Misalignment [m]	Range Misalignment [m]	Height Error [m]
CR 1	0.36	0.72	0.32	0.21	−0.14
CR 2	0.36	0.74	0.43	0.27	0.66
CR 3	0.36	0.80	0.42	0.26	−0.15
CR 4	0.38	0.74	0.37	0.42	0.65
CR 5	0.36	0.70	0.41	0.15	0.01
μ	0.36	0.74	0.39	0.26	0.20
σ	0.008	0.04	0.04	0.10	0.41

μ and σ indicate the mean and the standard deviation, respectively, of the achieved results.

As a second experiment, we measured the planar (that is, in the azimuth and range directions) positioning accuracy of the Single Look Complex (SLC) image in correspondence with the five CRs. To do this, we applied the backward geocoding procedure [1] to the D-GPS positions of the CRs, thus calculating their expected azimuth and range coordinates in the considered SAR image output grid. Then, we compared these coordinates with those of the CRs imaged in the SLC image. The range and azimuth misalignments measured for all the CRs are listed in Table 6. In particular, we measured a mean azimuth misalignment of 0.39 m with a standard deviation of 0.04 m, and a mean range misalignment of 0.26 m with a standard deviation of 0.10 m. It is noted that in the range direction the measured mean misalignment is lower than the resolution, whereas in the azimuth direction it is comparable to the resolution.

As a third experiment, we provided a first estimate of the vertical accuracy of the AXIS InSAR DEM. To do this, we compared the height values achieved on the AXIS InSAR DEM, in correspondence with the CR positions, with the D-GPS ones. The results are again collected in Table 6. In particular, we measured a mean vertical error of 0.20 m with a standard deviation of 0.41 m.

4. Discussion

A discussion concerning the presented results is now addressed.

In the previous section, to carry out the analysis of the AXIS performances, we have exploited five CRs properly deployed within the illuminated area and the D-GPS surveys of their positions. In this way, we have obtained a number of ground measurements that may be considered with good approximation as the ground truth.

In correspondence with this set of ground reference points we first measured the geometric resolution of the focused AXIS images. In particular, for the SLC image we have obtained mean geometric resolutions of 0.36 m (azimuth) × 0.74 m (range), with standard deviations of 0.008 m (azimuth) and 0.04 m (range). These values turned out to be practically the same as the expected theoretical ones.

Then, we measured the planar (that is, in the azimuth and range directions) positioning accuracy of the focused AXIS images, obtaining for the SLC image a mean positioning misalignment of 0.39 m (azimuth) × 0.26 m (range) with standard deviations of 0.04 m (azimuth) and 0.10 m (range). Finally, we compared the D-GPS measurements of the positions of the CRs with the height values of the single-pass InSAR AXIS DEM in correspondence with the imaged CRs, obtaining a mean vertical error of 0.20 m with a standard deviation of 0.41 m. From these results it turns out that some systematic errors seem to affect the obtained measurements. These errors are, however, tolerable, since the mean range misalignment (0.26 m) is less than the range resolution and spacing (0.75 m) of the system; the mean azimuth misalignment (0.39 m) is comparable to the azimuth resolution (0.33 m) of the system; and the mean vertical error is just 0.20 m. It is likely that these three systematic effects are somehow related. Given the amount of the mean range bias, one possible explanation could be the presence of an uncompensated internal radar delay. In any case, further investigations on this issue are a matter of current work.

Summing up, the measured imaging and topographic mapping capabilities of the overall AXIS infrastructure (which consists of the radar system along with the complete data processing chain that leads from the acquired raw data to the generated SAR and InSAR products) well match the theoretical, expected ones.

Moreover, the imaging and topographic mapping capabilities of this low cost, compact and flexible FMCW system are comparable to those achieved using well-assessed, although more expensive, pulsed airborne X-band SAR systems, like, for instance, InSAeS4 [9]. Indeed, in [9], starting from an analysis similar to that addressed in this work for the AXIS system, it was measured for InSAeS4 a mean geometric resolution of 0.14 m (azimuth) × 0.49 m (range), a mean positioning misalignment of 0.08 m (azimuth) × 0.04 m (range) with a standard deviation of 0.07 m (azimuth) and 0.08 m (range), and a mean height error (of the obtained single-pass InSAR DEM) of −0.08 m with a standard deviation of 0.51 m.

5. Conclusions

In this work, we presented a first assessment of the imaging and topographic mapping capabilities of the AXIS system, which is a newborn single-pass interferometric airborne FMCW SAR system developed in the frame of cooperation between a public research institute (IREA-CNR) and a private company (Elettra Microwave S.r.l.).

In particular, we showed results relevant to an acquisition campaign carried out over the Salerno area, South of Italy, in 2018, just after the system completion. More specifically, we provided a first quantitative assessment of the quality of the focused SAR images and InSAR products achieved using the AXIS system. To do this, we exploited a number of CRs properly deployed within the area illuminated by the radar and D-GPS surveys of their positions.

More specifically, in correspondence with this set of ground reference points we measured the geometric resolution and the positioning misalignment of the focused AXIS images, obtaining for the SLC image, mean geometric resolutions practically equal to the expected theoretical ones, a mean range misalignment smaller than the resolution and a mean azimuth misalignment comparable to the resolution. Moreover, we compared the D-GPS measurements of the positions of the CRs with the height values of the single-pass InSAR AXIS DEM in correspondence with the imaged CRs, obtaining a vertical error on the order of dozens of centimeters.

The presented results, aimed at providing first reference values for future research and operational activities that will be conducted with this system, already show that the imaging and topographic mapping capabilities of the AXIS system well match the theoretical, expected ones. Moreover, they are comparable to those achievable using well-assessed, although typically more expensive, pulsed SAR systems. More generally, the presented results show that the AXIS infrastructure (which consists of the radar system along with the complete data processing chain from the acquired raw data to the generated InSAR products) may represent an appealing monitoring solution for those applications that require the use of high-resolution InSAR products.

Author Contributions: Conceptualization, C.E., A.N., G.P., P.B., R.L., S.P.; Methodology, Software and Validation, C.E., A.N., P.B., S.P.; Investigation, C.E., A.N., G.P., P.B., R.L., S.P.; Data Curation, C.E., A.N., G.P., P.B., S.P.; Writing—Original Draft Preparation, C.E., A.N., S.P.; Supervision, S.P. All authors have read and agreed to the published version of the manuscript.

Funding: This work was supported in part by the Italian Department of Civil Protection (DPC) in the framework of the DPC-IREA Agreement (2019-2021) and in part by the Italian Ministry of Economic Development (MISE) in the framework of the contract "MATRAKA". Part of the presented research has been carried out through the I-AMICA (Infrastructure of High Technology for Environmental and Climate Monitoring-PONa3_00363).

Acknowledgments: The authors thank S. Guarino, F. Parisi and M. C. Rasulo of IREA-CNR, for their support, Etabeta srl for their contribution during the ground campaign, A. Gifuni of Università degli Studi di Napoli "Parthenope" for his contribution to the antennas measurements.

Conflicts of Interest: The authors declare no conflict of interest.

References

1. Franceschetti, G.; Lanari, R. *Synthetic Aperture Radar Processing*; CRC PRESS: New York, NY, USA, 1999.

2. Moreira, A.; Prats-Iraola, P.; Younis, M.; Krieger, G.; Hajnsek, I.; Papathanassiou, K.P. A tutorial on synthetic aperture radar. *IEEE Geosci. Remote Sens. Mag.* **2013**, *1*, 6–43. [CrossRef]

3. Krieger, G.; Moreira, A.; Fiedler, H.; Hajnsek, I.; Werner, M.; Younis, M.; Zink, M. TanDEM-X: A Satellite Formation for High-Resolution SAR Interferometry. *IEEE Trans. Geosci. Remote Sens.* **2007**, *45*, 3317–3341. [CrossRef]

4. Torre, A.; Calabrese, D.; Porfilio, M. COSMO-SkyMed: Image quality achievements. In Proceedings of the 5th International Conference on Recent Advances in Space Technologies—RAST2011, Istanbal, Turkey, 9–11 June 2011.

5. Torres, R.; Snoeij, P.; Geudtner, D.; Bibby, D.; Davidson, M.; Attema, E.; Potin, P.; Rommen, B.; Floury, N.; Brown, M.; et al. GMES Sentinel-1 Mission. *Remote Sens. Environ.* **2012**, *120*, 9–24. [CrossRef]

6. Le Toan, T.; Quegan, S.; Davidson, M.W.J.; Balzter, H.; Paillou, P.; Papathanassiou, K.; Plummer, S.; Rocca, F.; Saatchi, S.; Shugart, H.; et al. The BIOMASS mission: Mapping global forest biomass to better understand the terrestrial carbon cycle. *Remote Sens. Environ.* **2011**, *115*, 2850–2860. [CrossRef]

7. Kim, D.-J.; Hensley, S.; Yun, S.-H.; Neumann, M. Detection of Durable and Permanent Changes in Urban Areas Using Multitemporal Polarimetric UAVSAR Data. *IEEE Geosci. Remote Sens. Lett.* **2016**, *13*, 267–271. [CrossRef]

8. Baqué, R.; Bonin, G.; du Plessis, O.R. The airborne SAR-system: SETHI airborne microwave remote sensing imaging system. In Proceedings of the 7th European Conference on Synthetic Aperture Radar, Friedrichshafen, Germany, 2–5 June 2008.

9. Perna, S.; Esposito, C.; Amaral, T.; Berardino, P.; Jackson, G.; Moreira, J.; Pauciullo, A.; Vaz Junior, E.; Wimmer, C.; Lanari, R. The InSAeS4 airborne X-band interferometric SAR system: A first assesment on its imaging and topographic mapping capabilities. *Remote Sens.* **2016**, *8*, 40. [CrossRef]

10. Pinheiro, M.; Reigber, A.; Scheiber, R.; Prats-Iraola, P.; Moreira, A. Generation of highly accurate DEMs over flat areas by means of dual-frequency and dual-baseline airborne SAR interferometry. *IEEE Trans. Geosci. Remote Sens.* **2018**, *56*, 4361–4390. [CrossRef]

11. Perna, S.; Alberti, G.; Berardino, P.; Bruzzone, L.; Califano, D.; Catapano, I.; Ciofaniello, L.; Donini, E.; Esposito, C.; Facchinetti, C.; et al. The ASI Integrated Sounder-SAR System Operating in the UHF-VHF Bands: First Results of the 2018 Helicopter-Borne Morocco Desert Campaign. *Remote Sens.* **2019**, *11*, 1845. [CrossRef]

12. Aguasca, A.; Acevo-Herrera, R.; Broquetas, A.; Mallorqui, J.J.; Fabregas, X. ARBRES: Light-Weight CW/FM SAR Sensors for Small UAVs. *Sensors* **2013**, *13*, 3204–3216. [CrossRef]

13. Ouchi, K. Recent Trend and Advance of Synthetic Aperture Radar with Selected Topics. *Remote Sens.* **2013**, *5*, 716–807. [CrossRef]

14. Brenner, A.R.; Roessing, L. Radar Imaging of Urban Areas by Means of Very High-Resolution SAR and Interferometric SAR. *IEEE Trans. Geosci. Remote Sens.* **2008**, *46*, 2971–2982. [CrossRef]

15. Dubois-Fernandez, P.; du Plessis, O.R.; le Coz, D.; Dupas, J.; Vaizan, B.; Dupuis, X.; Cantalloube, H.; Coloumbeix, C.; Titin-Schnaider, C.; Dreuillet, P.; et al. The ONERA RAMSES SAR system. In Proceedings of the International Geoscience and Remote Sensing Symposium, Toronto, ON, CA, 24–28 June 2002.

16. Ruault du Plessis, O.; Nouvel, J.; Baque, R.; Bonin, G.; Dreuillet, P.; Coulombaix, C.; Oriot, H. ONERA SAR facilities. *IEEE Aerosp. Electron. Syst. Mag.* **2011**, *26*, 24–30. [CrossRef]

17. Perna, S.; Wimmer, C.; Moreira, J.; Fornaro, G. X-Band Airborne Differential Interferometry: Results of the OrbiSAR Campaign Over the Perugia Area. *IEEE Trans. Geosci. Remote Sens.* **2008**, *46*, 489–503. [CrossRef]

18. Magnard, C.; Frioud, M.; Small, D.; Brehm, T.; Essen, H.; Meier, E. Processing of MEMPHIS Ka-Band Multibaseline Interferometric SAR Data: From Raw Data to Digital Surface Models. *IEEE J. Sel. Top. Appl. Earth Observ. Remote Sens.* **2014**, *7*, 2927–2941. [CrossRef]

19. Schmitt, M.; Stilla, U. Maximum-likelihood based approach for single-pass synthetic aperture radar tomography over urban areas. *IET Radar Sonar Navig.* **2014**, *8*, 1145–1153. [CrossRef]

20. Fornaro, G.; Franceschetti, G.; Perna, S. Motion Compensation Errors. Effects on the Accuracy of Airborne SAR Images. *IEEE Trans. Aerosp. Electron. Syst.* **2005**, *41*, 1338–1352. [CrossRef]

21. Tomiyasu, K. Conceptual Performance of a Satellite Borne, Wide Swath Synthetic Aperture Radar. *IEEE Trans. Geosci. Remote Sens.* **1981**, *2*, 108–116. [CrossRef]

22. Gebert, N.; Krieger, G.; Moreira, A. Multichannel azimuth processing in ScanSAR and TOPS mode operation. *IEEE Trans. Geosci. Remote Sens.* **2010**, *48*, 2994–3008. [CrossRef]

23. De Zan, F.; Monti Guarnieri, A. TOPSAR: Terrain Observation by Progressive Scans. *IEEE Trans. Geosci. Remote Sens.* **2006**, *44*, 2352–2360. [CrossRef]

24. Gebert, N.; Krieger, G.; Moreira, A. Digital beamforming on receive: Techniques and optimization strategies for high-resolution wide-swath SAR imaging. *IEEE Trans. Aerosp. Electron. Syst.* **2009**, *45*, 564–592. [CrossRef]

25. Anantrasirichai, N.; Biggs, J.; Albino, F.; Hill, P.; Bull, D. Application of Machine Learning to Classification of Volcanic Deformation in Routinely Generated InSAR Data. *J. Geophys. Res. Solid Earth* **2018**, *123*, 6592–6606. [CrossRef]

26. Ma, P.; Zhang, F.; Lin, H. Prediction of InSAR time-series deformation using deep convolutional neural networks. *Remote Sens. Lett.* **2019**, *11*, 137–145. [CrossRef]

27. Anantrasirichai, N.; Biggs, J.; Albino, F.; Bull, D. A deep learning approach to detecting volcano deformation from satellite imagery using synthetic datasets. *Remote Sens. Environ.* **2019**, *230*, 111–179. [CrossRef]

28. Valade, S.; Ley, A.; Massimetti, F.; D'Hondt, O.; Laiolo, M.; Coppola, D.; Loibl, D.; Hellwich, O.; Walter, T.R. Towards Global Volcano Monitoring Using Multisensor Sentinel Missions and Artificial Intelligence: The MOUNTS Monitoring System. *Remote Sens.* **2019**, *11*, 1528. [CrossRef]

29. Richards, M.A.; Scheer, J.A.; Holm, W.A. *Principles of Modern Radar: Basic Principles*; SciTech Publishing: Raleigh, NC, USA, 2010.

30. Meta, A.; Hoogeboom, P.; Ligthart, L.P. Signal Processing for FMCW SAR. *IEEE Trans. Geosci. Remote Sens.* **2007**, *45*, 3519–3532. [CrossRef]

31. Balanis, C.A. *Antenna Theory: Analysis and Design*, 3rd ed.; Wiley-Interscience: New York, NY, USA, 2005.

32. Esposito, C.; Gifuni, A.; Perna, S. Measurement of the Antenna Phase Center Position in Anechoic Chamber. *IEEE Antennas Wirel. Propag. Lett.* **2018**, *17*, 2183–2187. [CrossRef]

33. Wimmer, C.; Siegmund, R.; Schwabisch, M.; Moreira, J. Generation of high precision DEMs of the Wadden Sea with airborne interferometric SAR. *IEEE Trans. Geosci. Remote Sens.* **2000**, *38*, 2234–2245. [CrossRef]

34. Esposito, C.; Natale, A.; Palmese, G.; Berardino, P.; Perna, S. Geometric distortions in FMCW SAR images due to inaccurate knowledge of electronic radar parameters: Analysis and correction by means of corner reflectors. *Remote Sens. Environ.* **2019**, *232*, 111289. [CrossRef]

35. Berardino, P.; Esposito, C.; Natale, A.; Lanari, R.; Perna, S. Airborne SAR Focusing in the Presence of Severe Squint Variations. In Proceedings of the IGARSS 2019—2019 IEEE International Geoscience and Remote Sensing Symposium, Yokohama, Japan, 28 July–2 August 2019; pp. 537–540.

36. Frey, O.; Magnard, C.; Ruegg, M.; Meier, E. Focusing of airborne Synthetic Aperture Radar data from highly nonlinear flight tracks. *IEEE Trans. Geosci. Remote Sens.* **2009**, *47*, 1844–1858. [CrossRef]

37. Rabus, B.; Eineder, M.; Roth, A.; Bamler, R. The shuttle radar topography mission—A new class of digital elevation models acquired by spaceborne radar. *ISPRS J. Photogramm. Remote Sens.* **2003**, *57*, 241–262. [CrossRef]

38. Fornaro, G.; Franceschetti, G.; Perna, S. On Center-Beam Approximation in SAR Motion Compensation. *IEEE Geosci. Remote Sens. Lett.* **2006**, *3*, 276–280. [CrossRef]

39. Moreira, A.; Huang, Y. Airborne SAR Processing of Highly Squinted Data Using a Chirp Scaling Approach with Integrated Motion Compensation. *IEEE Trans. Geosci. Remote Sens.* **1994**, *32*, 1029–1040. [CrossRef]

40. Fornaro, G. Trajectory Deviations in Airborne SAR: Analysis and Compensation. *IEEE Trans. Aerosp. Electron. Syst.* **1999**, *35*, 997–1009. [CrossRef]

41. Costantini, M. A novel phase unwrapping method based on network programming. *IEEE Trans. Geosci. Remote Sens.* **1998**, *36*, 813–821. [CrossRef]

42. Dudczyk, J.; Kawalec, A. Optimizing the minimum cost flow algorithm for the phase unwrapping process in SAR radar. *Bull. Pol. Acad. Sci. Tech. Sci.* **2014**, *62*, 511–516. [CrossRef]

43. Esposito, C.; Pauciullo, A.; Berardino, P.; Lanari, R.; Perna, S. A Simple Solution for the Phase Offset Estimation of Airborne SAR Interferograms Without Using Corner Reflectors. *IEEE Geosci. Remote Sens. Lett.* **2017**, *14*, 379–383. [CrossRef]

44. Perna, S.; Esposito, C.; Berardino, P.; Pauciullo, A.; Wimmer, C.; Lanari, R. Phase Offset Calculation for Airborne InSAR DEM Generation Without Corner Reflectors. *IEEE Trans. Geosci. Remote Sens.* **2014**, *53*, 2713–2726. [CrossRef]

45. Fornaro, G.; Sansosti, E.; Lanari, R.; Tesauro, M. Role of processing geometry in SAR raw data focusing. *IEEE Trans. Aerosp. Electron. Syst.* **2002**, *38*, 441–454. [CrossRef]

Letter

Drone-Borne Differential SAR Interferometry

Dieter Luebeck [1], Christian Wimmer [2], Laila F. Moreira [1,3], Marlon Alcântara [3], Gian Oré [3], Juliana A. Góes [3], Luciano P. Oliveira [3,*], Bárbara Teruel [4], Leonardo S. Bins [5], Lucas H. Gabrielli [3] and Hugo E. Hernandez-Figueroa [3]

[1]   Radaz Indústria e Comércio de Produtos Eletrônicos Ltda., São José dos Campos, José dos Campos 12244-000, Brazil; dieter.luebeck@t-jump.net (D.L.); laila.moreira@radaz.com.br (L.F.M.)
[2]   Wimmer Consulting, 84061 Ergoldsbach, Germany; office@wimmer-christian.de
[3]   School of Electrical and Computer Engineering, University of Campinas-UNICAMP, Campinas 13083-852, Brazil; marlon.s.alcantara@outlook.com.br (M.A.); gc.ore.huacles@gmail.com (G.O.); juliana.goes@engineer.com (J.A.G.); lucashg@fee.unicamp.br (L.H.G.); hugoehf@gmail.com (H.E.H.-F.)
[4]   School of Agricultural Engineering, University of Campinas-UNICAMP, Campinas 13083-875, Brazil; barbara.teruel@feagri.unicamp.br
[5]   National Institute for Space Research–INPE, São José dos Campos 12227-010, Brazil; leonardo.bins@gmail.com
*   Correspondence: luciano@decom.fee.unicamp.br; Tel.: +55-19-99144-3710

Received: 31 December 2019; Accepted: 17 February 2020; Published: 29 February 2020

Abstract: Differential synthetic aperture radar interferometry (DInSAR) has been widely applied since the pioneering space-borne experiment in 1989, and subsequently with the launch of the ERS 1 program in 1992. The DInSAR technique is well assessed in the case of space-borne SAR data, whereas in the case of data acquired from aerial platforms, such as airplanes, helicopters, and drones, the effective application of this technique is still a challenging task, mainly due to the limited accuracy of the information provided by the navigation systems mounted onboard the platforms. The first airborne DInSAR results for measuring ground displacement appeared in 2003 using L- and X-bands. DInSAR displacement results with long correlation time in P-band were published in 2011. This letter presents a SAR system and, to the best of our knowledge, the first accuracy assessment of the DInSAR technique using a drone-borne SAR in L-band. A deformation map is shown, and the accuracy and resolution of the methodology are presented and discussed. In particular, we have obtained an accuracy better than 1 cm for the measurement of the observed ground displacement. It is in the same order as that achieved with space-borne systems in C- and X-bands and the airborne systems in X-band. However, compared to these systems, we use here a much longer wavelength. Moreover, compared to the satellite experiments available in the literature and aimed at assessing the accuracy of the DInSAR technique, we use only two flight tracks with low time decorrelation effects and not a big data stack, which helps in reducing the atmospheric effects.

Keywords: differential interferometry; DInSAR; drone-borne radar; range-Doppler processor; corner reflector

---

1. Introduction

Differential synthetic aperture radar interferometry (DInSAR) became popular after the launch of the ERS 1 satellite in 1991. Relevant satellite-borne DInSAR results already appeared in 1989 when Gabriel et al. [1] made two interferograms from three SAR images, taken from the same area at different times. They produced a double-difference interferogram to remove phase shifts caused by topography and to retain phase changes due to surface motion; they were able to measure surface motions of up to 1 cm with a 10-m resolution over 50 km swaths. Since then, a large number of deformation maps, generated from satellite-borne DInSAR data, have been supporting many geological and engineering monitoring tasks and studies. Nonetheless, satellite-borne DInSAR systems present an operational limitation due to the fixed orbits of the satellites, as DInSAR only measures displacement in the line-of-sight direction, and to retrieve a 3-D deformation map, information from at least three viewing angles is required. Some methods accomplish this by acquiring quasi-simultaneous SAR images with different squint angles [2,3]. Other methods use multi-interferograms from more than three directions [4] or the multiple aperture interferometry technique [5,6], which measures displacement in the along-track direction.

Airborne DInSAR for displacement measurements was first presented in 2003 using L-band [7], in 2004 using X-band [8], and in 2011 using P-band [9]. The first report of a controlled accuracy measurement using corner reflectors and distributed targets on the ground was published in 2008 [10]. Since aircraft have flexible flight patterns, surveying the area in three different directions is not a complicated task; therefore, generating 3-D deformation maps with airborne DInSAR is more straightforward than with satellite-borne systems. Moreover, brief revisit periods with high spatial resolution can be achieved. The airborne DInSAR solution has been implemented with limitation due to the high survey costs and the smaller illumination swath width, compared with satellite-borne systems. Additionally, aircraft movement caused by air turbulence is a hindrance to accurately executing airborne DInSAR, and hence many motion compensation techniques have been applied. Instead, Cao et al. [11] used the time-domain back-projection algorithm for SAR processing, which is intrinsically able to balance out platform motion. They demonstrated that it is possible to measure deformation with centimeter accuracy.

A novel drone-borne SAR system operating in the P-, L-, and C-bands and optimized for the DInSAR operation was presented in 2019 [12]. This new SAR system has a much smaller swath width when compared with airborne DInSAR systems and can cover areas of about tens of square kilometers. Thus, small-size drone-borne DInSAR systems are suitable for applications never attended by satellite and airborne DInSAR. A first DInSAR demonstration is introduced in [12], which had an RMS displacement measurement error of 0.045 m in L-band.

Deformation maps accurate to the order of a few millimeters are very useful as input to early warning systems for landslide and dam failure processes, or emergency management systems after earthquakes. The five most important features for having good differential interferometry results are high signal-to-noise ratio, low time decorrelation between the revisits, high radar and processing phase stability, interference-free antennae environment, and precise motion data. Each of these points will be quantitatively addressed.

This letter reports the first results of the drone-borne DInSAR obtained from a controlled experiment with a set of corner reflectors. During each of three subsequent flights, one of the corner reflectors was displaced, while the others remained fixed. The displacement measurement was six times more accurate than that presented in [12].

2. Materials and Methods

The radar prototype described in [12] operates in three bands: P, L, and C. Also, five channels are available: two C-band antennae for cross-track interferometry, two L-band antennae for polarimetry, and one P-band antenna. The results presented in this letter focus on the L-band HH-polarization for the DInSAR, in which the radar transmits and receives horizontally polarized waves. Table 1 shows the main radar acquisition parameters.

The digital surface model (DSM) is determined by using the cross-track interferometry information provided by the two C-band antennae. Then, the DSM is applied in the DInSAR calculation. The nominal height accuracy is better than 1 m RMS with a spatial resolution of 2 m.

The motion sensing system (MSS) is integrated into the radar, as shown in [12], and it consists of a single channel GNSS receiver and an inertial measurement unit (IMU). One ground station with a single channel GNSS receiver is also utilized to allow differential GNSS processing. Table 2 shows the MSS requirements [12].

Table 1. Main radar acquisition parameters.

Radar Parameters	Value
Carrier Wavelength	22.84 cm
Bandwidth	150 MHz
Polarization	HH
Peak Power	100 mW
Mean Power	1 mW
PRF	10 kHz
Incidence Angle	45 deg.
Mean Drone Height	120 m
Mean Drone Velocity	2 m/s
Motion Sensing System (MSS)	D-GNSS+IMU
Range Resolution	1 m
Azimuth Resolution	0.1 m
Processed Azimuth Bandwidth	20 Hz
Processed Aperture at 45 deg. Incidence Angle	196 m
SLC Range Sampling	0.61 m
SLC Azimuth Sampling	0.05 m

Table 2. Accuracy specifications of the adopted motion sensing system (MSS) [12].

MSS Accuracy Specifications	Value
Position, Absolute	0.015 m
Position, Relative	<0.01 m
Roll and Pitch	0.1 deg.
True Heading	2 deg.
IMU Run Bias Stability	2 deg./h
IMU Angular Random Walk	0.1 deg./square root hour

This letter shows the tests and the results of an L-band drone-borne DInSAR survey carried out at the University of Campinas, UNICAMP, Campinas, Brazil. Three trihedral corner reflectors with square sides and an edge length of 0.6 m were used as a ground reference. The resulting radar cross-section is 20 dB. Considering a 1-square-meter resolution cell, the respective sigma zero is 20 dB as the ground reflectivity has a sigma zero of about −5 dB.

The experiment with the drone-borne system, shown in Figure 1a, took place on 19 December 2018 at the UNICAMP test site, as follows.

- The GNSS ground station was placed close to the starting position of the drone, and the GNSS recording was initiated.
- Three flights were carried out, each consisting of the following successive steps: turning on the drone and the radar; waiting 15 min for simultaneous and stationary recording of ground station and radar GNSS data; take off; executing the same west-east flight track; landing; waiting 15 min for simultaneous and stationary recording of ground station and radar GNSS data; turning-off the radar and the drone.
- The three corner reflectors were positioned at the UNICAMP test site at different radar incidence angles, between 36 and 60 degrees, as shown in Figure 2. Corner reflector 1, in the nearest range, had an incidence angle of 36 degrees. Corner reflector 2, in the middle range between 1 and 3, had an incidence angle of 52 degrees. Corner reflector 3, in far range, had an incidence angle of 60 degrees. Corner 2 was lifted differently during the second and third flights to validate the displacement measurement methodology.
- During the first flight, all three corner reflectors were placed horizontally on the ground, looking north.

- During the second flight, only the corner reflector number 2 was lifted 4 cm off the ground, as shown in Figure 1b. Corner reflectors 1 and 3 remained at ground level.
- During the third flight, only the corner reflector number 2 was lifted 2 cm off the ground. Corner reflectors 1 and 3 were still at ground level.
- The GNSS ground station and the drone were dismounted. The acquired data were downloaded for processing.

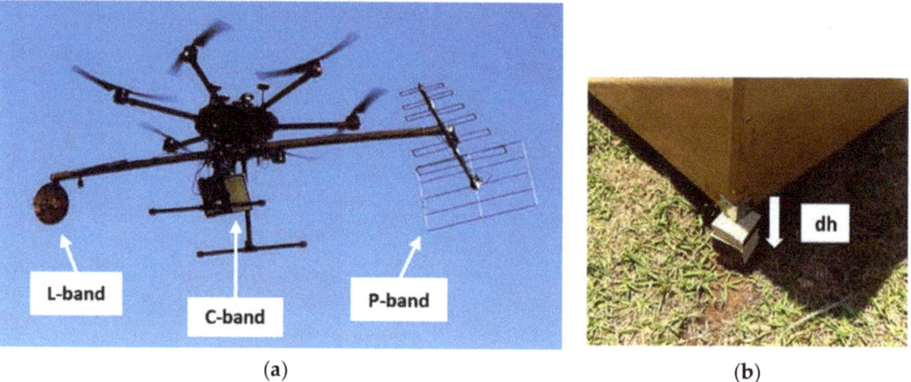

(a) (b)

Figure 1. Experiment components. (**a**) Drone-borne differential synthetic aperture radar interferometry (DInSAR) system [12] including one L-band antenna with H and V dipoles; two interferometric C-band antennae attached to the radar electronics; one P-band antenna with an H dipole. (**b**) Corner reflector 2, which was lifted 4 cm off the ground during the second flight and 2 cm during the third flight. Displacement is positive (dh > 0) when terrain height decreases with time.

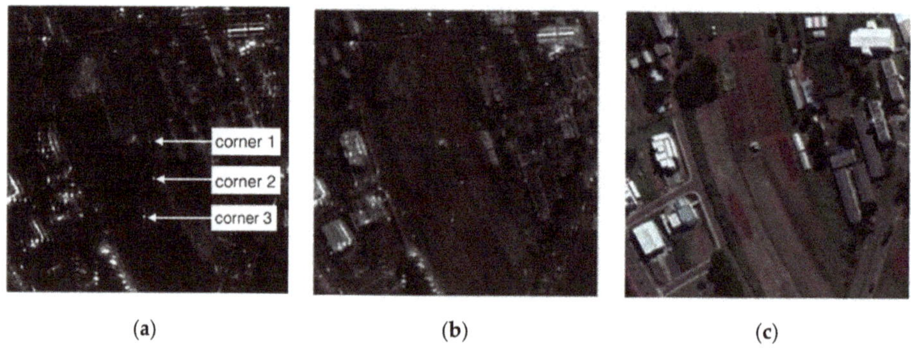

(a) (b) (c)

Figure 2. Test area at UNICAMP, Campinas Brazil. (**a**) Geocoded L-band SAR image of the first flight, with the indication of the position of the three corner reflectors: corner 1 in the nearest range, corner 2 in the middle range, and corner 3 in the far range. Flight track on the top of the image, from west to east (left to right), looking south (downward). (**b**) Overlay of SAR and optical images to show the target's position. (**c**) Optical image taken on the same day.

After the three flights, the data were processed as follows:

1. Perform the differential GNSS processing of ground station and radar GNSS receivers.
2. Generate position and antennae orientation history with IMU and differential GNSS data fusion.
3. Process the raw radar data for each of the three flight paths mentioned above (raw data 1, 2, and 3) according to the processing chain described in [10] and shown in Figure 3: range and azimuth

compression with motion compensation; recording of the three resulting single-look complex (SLC) images; interferometry (SLC 1 with SLC 2, generating the interferogram 1 and 2, and SLC 1 with SLC 3, producing the interferogram 1 and 3); topography subtraction and phase to height conversion; and complex filtering. The output consists of two deformation maps, plus the three SLCs, all in slant range geometry. Each interferogram is calculated from the SLCs with 0.1-m resolution in azimuth and 1 m resolution in the slant range. Complex filtering results on a final spatial resolution of 3 m in both slant range and azimuth directions for each deformation map.

4. Check the absolute position of the corner reflectors by searching their maximum power positions on the three slant range SLC images.

5. Use the azimuth and slant range coordinates found in step 4 to read the displacement for each corner reflector on the deformation maps. Compare the displacement with the expected results:

   a. Null differential height offsets for corner reflectors 1 and 3 on both deformation maps, since they were always at ground level;

   b. A 4 cm lift for corner reflector 2 on the deformation map of interferogram 1 and 2;

   c. A 2 cm lift for corner reflector 2 on the deformation map of interferogram 1 and 3.

6. Geocode the two deformation maps.

7. Generate the multi-look images from the SLCs with 1 m × 1 m resolution in slant range and geocode them, as shown in Figure 2a.

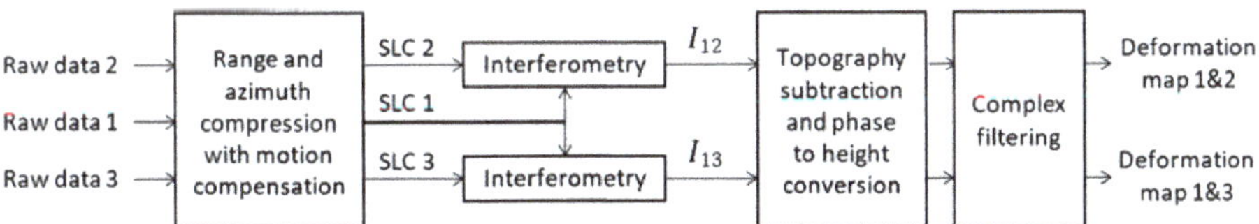

Figure 3. Simplified flow chart of the processing chain. The complete version is in [10]. I_{12} is the interferogram 1 and 2, and I_{13} is the interferogram 1 and 3.

The chosen processing chain [10] was originally applied to an airborne DInSAR experiment. Nevertheless, it can be extended to this case thanks to the stability of the drone, which will be addressed in the next section. In particular, azimuth compression can be done at Zero Doppler, and motion compensation algorithms can be as effective as in the case of airborne DInSAR. The processing algorithm described in [12] was not used for this experiment because, unlike the processing chain reported in [10], it does not consider the topography subtraction step and does not deal with residual motion errors, which cause phase errors in the slant range SLC images.

3. Results

The drone performed three stable flights presenting an orientation error of less than 1-degree RMS and a position error of less than 0.2 m RMS. This behavior represented the majority of the research flights and was a positive surprise: well-stabilized hexacopter drones can repeat the same track within a 50 cm diameter tube and can hold the heading very close to the course direction. Figure 4 presents the deformation map derived from the first and the second flights, and Figure 5 shows the deformation map derived from the first and the third flights. Both Figures 4 and 5 are geocoded and overlaid on an optical image. The displacement was measured according to steps 4 and 5 of the last section

Figure 4. Deformation map of the interferogram 1 and 2—from the first and the second flights. Displacement is positive (dh > 0) when terrain height decreases with time. The coherence threshold is 0.6. Corner reflector 2 was lifted 4 cm in the second flight. The measured displacement was −45 mm.

Figure 5. Deformation map of the interferogram 1 and 3—from the first and the third flights. Displacement is positive (dh > 0) when terrain height decreases with time. The coherence threshold is 0.75. Corner reflector 2 was lifted 2 cm in the second flight. The measured displacement was −15 mm.

Table 3 shows the results overview of the two deformation maps. The displacement measurement error range is about 10 mm. The overall standard deviation of the displacement measurement is 7.4 mm, calculated with the six height difference values from the three corner reflectors and both deformation maps. It is six times better than the 45 mm error reported in [12]. Also, surfaces with a coherence better than 0.6, excluding the area of the corner reflector 2, show a standard deviation between 4 and 6 mm, as presented in Table 4.

Table 3. Results overview of the two deformation maps.

Corner Reflectors	Deformation Map of 1 and 2 [mm]			Deformation Map 1 and 3 [mm]		
	Ideal	Measured	Difference	Ideal	Measured	Difference
1	0	11	11	0	5	5
2	−40	−45	−5	−20	−15	5
3	0	−6	−6	0	−10	−10

Table 4. Deformation map error overview, considering both deformation maps—right column: standard deviation of the deformation map error.

Targets	Standard Deviation [mm]
Corner Reflectors	7.4
Areas with Coherence Better than 0.6	4–6

The analysis of the results aforementioned is structured in the following five aspects mentioned in Section 1:

1. signal-to-noise ratio;
2. time decorrelation between the revisits;
3. radar and processing phase stability;
4. antennae environment;
5. motion data.

3.1. Signal-to-Noise Ratio

The measured signal-to-noise ratio (SNR) of the corner reflectors is 32 dB, and their reflectivity is 20 dB. The SNR and the reflectivity of the surrounding grass area are 7 dB and −5 dB, respectively. The SNR of the neighboring bare soil is 3 dB, and the reflectivity is −9 dB, as shown in Table 5. The expected coherence, the unfiltered interferometric phase noise, the filtered phase noise, and the standard deviation of the displacement error are calculated according to [13]. This standard deviation is caused exclusively by the SNR. The filtered phase has a resolution of 3 m by 3 m, starting from the unfiltered phase with 1 m resolution in range and 0.1-m resolution in azimuth. The resulting standard deviation of the displacement measurement error is 0.13 mm for the corner reflectors, 1.6 mm for the grass or distributed targets with an average reflectivity of −5 dB, and 2.2 mm for bare soil or distributed targets with an average reflectivity of −9 dB.

Table 5. DInSAR parameters calculated considering the reflectivity and signal-to-noise ratio (SNR) of the corner reflector and the grass surface.

Targets	Reflectivity [dB]	SNR [dB]	Coherence [0–1]	Unfiltered Phase Noise [deg.]	Filtered Phase Noise [deg.]	Standard Deviation of the Displacement Measurement Error [mm]
Corner reflectors	20	32	~1	4	0.4	0.13
Grass	−5	7	0.8	52	5.2	1.6
Bare soil	−9	3	0.6	70	7	2.2

3.2. Time Decorrelation Between the Revisits

As the radar operates in the L-band, the decorrelation within a day is negligible [7]. The high coherence of the interferograms 1 and 2 and 1 and 3 also shows that the contribution of hours of the revisit time does not have any influence on the degradation of the deformation map accuracy. The measured coherence is compatible with the measured SNR and does not contribute the time decorrelation.

3.3. Radar and Processing Phase Stability

The phase errors caused by the radar and the processing chain are less than 1 degree. The resulting error has a standard deviation at the deformation map of less than 0.4 mm.

3.4. Antennae Environment

Airborne SAR systems have considerable phase errors caused by secondary reflections of the pulse transmission and backscatter signal reflections at the fuselage and wings. Thus, several corrections and considerations are required. Fortunately, this was not a great issue for the drone-borne DInSAR system, shown in Figure 1a, since the influence of the class 3 drone on the radar was negligible. No changes in the antenna diagram, no variation in the antenna gain, and no secondary reflections were observed.

3.5. Motion Data

Considering Table 2, the major contribution for the MSS error is from the differential GNSS calculation as the orientation errors are negligible due to the small lever arms between the MSS and the L-band antenna. For the DInSAR, only the relative position accuracy is relevant, as the absolute position error can be compensated by taking a ground reference target. A precise differential GNSS calculation with a fixed phase solution was obtained because the ground station was always closer than 2 km to the drone, and the integration time before and after each flight was greater than 10 min. The IMU/D-GNSS fusion program delivers only a rough estimation of the relative position accuracy. Here, a relative position accuracy was estimated at 5 mm.

4. Discussion

Table 6 presents an overview of the deformation map error sources. The total measured error corresponds to the geometric average of all five contributions mentioned in Sections 3.1–3.5, as all these sources are statistically uncorrelated. The error contributions from SNR, time decorrelation, radar and processing phase stability, and antenna environment can be neglected when compared to the motion data error contribution. The MSS processing error, estimated at 5 mm, is by far the main source of error for the deformation map accuracy since it is approximately equal to the total deformation map error of about 4–7.4 mm. Regarding the SNR, the breakdown of Table 5 shows that all three target types—the corner reflectors and the grass and bare soil areas—contribute less than the motion data error to the total measured error. Both distributed and point targets have the same relation between coherence and phase noise in the interferogram, due to the short revisit time that causes a negligible time decorrelation.

Table 6. Error source budget of the deformation map. There are five sources and a total measured error of about 4–7.4 mm.

Error Sources	Standard Deviation [mm]
SNR	<2.2
Time Decorrelation	~0
Radar and Processing Phase Stability	<0.4
Antenna Environment	~0
Motion Data	~5
Total Measured RMS Error	4–7.4

The coherence thresholds of Figures 4 and 5 are 0.6 and 0.75, respectively. If both deformation maps had the same threshold value, the two figures would be nearly identical, except for the area related to corner reflector 2. A comparison between them shows that the most appropriate coherence threshold for these interferograms is 0.6 since the use of 0.75 would discard much useful information. This is why all the quantitative results were presented for a coherence threshold of 0.6.

5. Conclusions

The drone-borne DInSAR system introduced in [12] was operated to perform the experiment presented in this work and was able to generate deformation maps with 4–7.4 mm accuracy. Ground deformation maps with such high accuracy are valuable for monitoring areas and constructions in early warning and emergency management systems for disaster mitigation. Furthermore, the easy transportability of the drone-borne DInSAR system can allow quick and low-cost monitoring of small areas like dams and mining sites.

An overview of the accuracy assessments of deformation maps can be summarized as follows.

1. The accuracy assessment in L-band presented here is, as far as we know, novel;
2. C- and X-band satellite-borne systems reported in [14–16] and the airborne system published in [10] have similar accuracy.

The achievement of a high and stable deformation map accuracy with only two flight tracks and with low time decorrelation effects reported here is, to the best of our knowledge, unique.

In future works, multiple frequency differential GNSS shall improve accuracy. However, the improvement will not be very significant as the range between the GNSS ground station and the drone will always be less than a few kilometers. Additionally, P- and C-bands deformation maps shall be explored and merged with the L-band one. A three-band deformation map will have improved accuracy and a decorrelation time of several years due to the P-band [9]. Additionally, we have been working with the time-domain back-projection algorithm and expect to publish DInSAR results in the near future.

Author Contributions: Investigation, methodology, software, and validation, D.L., C.W., L.F.M., M.A., and G.O.; formal analysis and writing, D.L., C.W., L.F.M., J.A.G., and L.P.O.; supervision, review, and funding acquisition, D.L., B.T., L.S.B., L.H.G., and H.E.H.-F. All authors have read and agreed to the published version of the manuscript.

Funding: This research was funded by government agencies CAPES and CNPq and the São Paulo State agency FAPESP under the contract PIPE 2018/00601-8 and PITE 2017/19416-3.

Acknowledgments: We thank the teams of the School of Electrical and Computer Engineering and the School of Agricultural Engineering of the University of Campinas, UNICAMP, for the support given.

Conflicts of Interest: The authors declare no conflict of interest.

References

1. Gabriel, A.K.; Goldstein, R.M.; Zebker, H.A. Mapping small elevation changes over large areas: Differential radar interferometry. *J. Geophys. Res.* **1989**, *94*, 9183. [CrossRef]
2. Prats-Iraola, P.; Lopez-Dekker, P.; De Zan, F.; Yagüe-Martínez, N.; Zonno, M.; Rodriguez-Cassola, M. Performance of 3-D surface deformation estimation for simultaneous squinted SAR acquisitions. *IEEE Trans. Geosci. Remote Sens.* **2018**, *56*, 2147–2158. [CrossRef]
3. Jung, H.; Lu, Z.; Sherhperd, A.; Wright, T. Simulation of the superSAR multi-azimuth Synthetic Aperture Radar imaging system for precise measurement of three-dimensional earth surface displacement. *IEEE Trans. Geosci. Remote Sens.* **2015**, *53*, 6196–6206. [CrossRef]
4. Kimura, H. Three-dimensional surface deformation mapping from multi directional SAR interferograms. In Proceedings of the 2017 IEEE International Geoscience and Remote Sensing Symposium (IGARSS), Fort Worth, TX, USA, 23–28 July 2017; pp. 1692–1695.
5. Jo, M.; Jung, H.; Won, J.; Poland, M.P.; Miklius, A. Measurement of three-dimensional surface deformation of the March 2011 Kamoamoa fissure eruption, Kilauea Volcano, Hawai'i. In Proceedings of the 2014

IEEE Geoscience and Remote Sensing Symposium (IGARSS), Quebec City, QC, Canada, 13–18 July 2014; pp. 437–440.

6. Jung, H.; Yung, S.; Jo, M. An improvement of Multiple-Aperture SAR interferometry performance in the presence of complex and large line-of-sight deformation. *IEEE J. Sel. Top. Appl. Earth Obs. Remote Sens.* **2015**, *8*, 1743–1752. [CrossRef]

7. Reigber, A.; Scheiber, R. Airborne differential SAR interferometry: First results at L-band. *IEEE Trans. Geosci. Remote Sens.* **2003**, *41*, 1516–1520. [CrossRef]

8. Fornaro, G.; Lanari, R.; Sansosti, E.; Franceschetti, G.; Perna, S.; Gois, A.; Moreira, J. Airborne differential interferometry: X-band experiments. In Proceedings of the 2004 IEEE International Geoscience and Remote Sensing Symposium (IGARSS), Anchorage, AK, USA, 20-24 September 2004; Volume 5, pp. 3329–3332.

9. De Macedo, K.A.C.; Wimmer, C.; Barreto, T.L.M.; Lübeck, D.; Moreira, J.R.; Rabaco, L.M.L.; de Oliveira, W.J. Long-term airborne DInSAR measurements: P- and X-band cases. In Proceedings of the 2011 IEEE International Geoscience and Remote Sensing Symposium (IGARSS), Vancouver, BC, Canada, 24-29 July 2011; pp. 1175–1178.

10. Perna, S.; Wimmer, C. X-Band Airborne Differential Interferometry: Results of the OrbiSAR Campaign Over the Perugia Area. *IEEE Trans. Geosci. Remote Sens.* **2008**, *46*, 489–503. [CrossRef]

11. Cao, N.; Lee, H.; Zaugg, E.; Shrestha, R.; Carter, W.; Glennie, C.; Wang, G.; Lu, Z.; Fernandez-Diaz, J.C. Airborne DInSAR Results Using Time-Domain Backprojection Algorithm: A Case Study Over the Slumgullion Landslide in Colorado With Validation Using Spaceborne SAR, Airborne LiDAR, and Ground-Based Observations. *IEEE J. Sel. Top. Appl. Earth Obs. Remote Sens.* **2017**, *10*, 4987–5000. [CrossRef]

12. Moreira, L.; Castro, F.; Góes, J.A.; Bins, L.; Teruel, B.; Fracarolli, J.; Castro, V.; Alcântara, M.; Oré, G.; Luebeck, D.; et al. A Drone-borne Multiband DInSAR: Results and Applications. In Proceedings of the 2019 IEEE Radar Conference (RadarConf), Boston, MA, USA, 22-26 April 2019; pp. 1–6.

13. Just, D.; Bamler, R. Phase statistics of interferograms with applications to synthetic aperture radar. *Appl. Opt.* **1994**, *33*, 4361. [CrossRef] [PubMed]

14. Casu, F.; Manzo, M.; Lanari, R. A quantitative assessment of the SBAS algorithm performance for surface deformation retrieval from DInSAR data. *Remote Sens. Env.* **2006**, *102*, 195–210. [CrossRef]

15. Bonano, M.; Manunta, M.; Pepe, A.; Paglia, L.; Lanari, R. From Previous C-Band to New X-Band SAR Systems: Assessment of the DInSAR Mapping Improvement for Deformation Time-Series Retrieval in Urban Areas. *IEEE Trans. Geosci. Remote Sens.* **2013**, *51*, 1973–1984. [CrossRef]

16. Manunta, M.; De Luca, C.; Zinno, I.; Casu, F.; Manzo, M.; Bonano, M.; Fusco, A.; Pepe, A.; Onorato, G.; Berardino, P.; et al. The Parallel SBAS Approach for Sentinel-1 Interferometric Wide Swath Deformation Time-Series Generation: Algorithm Description and Products Quality Assessment. *IEEE Trans. Geosci. Remote Sens.* **2019**, *57*, 6259–6281. [CrossRef]

Article

Crop Growth Monitoring with Drone-Borne DInSAR

Gian Oré [1], Marlon S. Alcântara [1], Juliana A. Góes [1], Luciano P. Oliveira [1,*], Jhonnatan Yepes [2], Bárbara Teruel [2], Valquíria Castro [1], Leonardo S. Bins [3], Felicio Castro [1], Dieter Luebeck [4], Laila F. Moreira [4], Lucas H. Gabrielli [1] and Hugo E. Hernandez-Figueroa [1]

[1] School of Electrical and Computer Engineering, University of Campinas—UNICAMP, Campinas 13083-852, Brazil; g228005@dac.unicamp.br (G.O.); m228004@dac.unicamp.br (M.S.A.); jugoes@decom.fee.unicamp.br (J.A.G.); v162627@g.unicamp.br (V.C.); f180540@g.unicamp.br (F.C.); lucashg@fee.unicamp.br (L.H.G.); hugo@decom.fee.unicamp.br (H.E.H.-F.)
[2] School of Agricultural Engineering, University of Campinas—UNICAMP, Campinas 13083-875, Brazil; jayepesg@unal.edu.co (J.Y.); barbara.teruel@feagri.unicamp.br (B.T.)
[3] National Institute for Space Research—INPE, São José dos Campos 12227-010, Brazil; leonardo.bins@inpe.br
[4] Radaz Indústria e Comércio de Produtos Eletrônicos Ltda., São José dos Campos 12244-000, Brazil; dieter.luebeck@radaz.com.br (D.L.); laila.moreira@radaz.com.br (L.F.M.)
* Correspondence: luciano@decom.fee.unicamp.br; Tel.: +55-19-99144-3710

Received: 31 December 2019; Accepted: 3 February 2020; Published: 12 February 2020

Abstract: Accurate, high-resolution maps of for crop growth monitoring are strongly needed by precision agriculture. The information source for such maps has been supplied by satellite-borne radars and optical sensors, and airborne and drone-borne optical sensors. This article presents a novel methodology for obtaining growth deficit maps with an accuracy down to 5 cm and a spatial resolution of 1 m, using differential synthetic aperture radar interferometry (DInSAR). Results are presented with measurements of a drone-borne DInSAR operating in three bands—P, L and C. The decorrelation time of L-band for coffee, sugar cane and corn, and the feasibility for growth deficit maps generation are discussed. A model is presented for evaluating the growth deficit of a corn crop in L-band, starting with 50 cm height. This work shows that the drone-borne DInSAR has potential as a complementary tool for precision agriculture.

Keywords: differential interferometry; DInSAR; precision agriculture; drone-borne radar; crop growth deficit map

---

1. Introduction

Agriculture has a vital role in the economic stability and social development of a country. Effective agricultural management is essential to reduce costs and increase production. The monitoring of crop growth shall be done continuously for accurate support of decision-making [1]. Remote sensing has been an important tool for soil and crop monitoring. Optical remote sensing is widely used; nonetheless, synthetic aperture radar (SAR) remote sensing does not depend on weather conditions or sunlight. Moreover, the radar wavelength is approximately one million times greater than the wavelength of optical systems and provides complementary information about the agriculture parameters.

Several studies have shown the SAR remote sensing capabilities in growth monitoring of various crops [2,3] and crop classification [4–6]. Crop growth monitoring has been explored by using techniques based on statistical analysis between radar backscattering and crop height [2,3,7], with the use of techniques like interferometric SAR (InSAR) [8], polarimetric decomposition [9], polarimetric interferometric SAR (Pol-InSAR) [10] and differential SAR Interferometry (DInSAR) [10]. The DInSAR methodology presents high accuracy and spatial resolution, as it takes advantage of the phase difference between images. The most popular application of the DInSAR is the estimation of subsidence maps with millimetric accuracy. Relevant DInSAR works using satellites appeared in 1989, in 2003, using aircraft, and in 2019, employing drone-borne systems [11].

This article reports a first-stage crop growth estimation based on SAR experimental data from several circular flight surveys carried out over a test area of the School of Agricultural Engineering of the University of Campinas (UNICAMP) in Campinas, Brazil. It also proposes a model to estimate the corn crop growth, considering different stages of crop phenology. The types of flight surveys are discussed, and a growth map for a corn crop is presented. The drone-borne system presented in [12] was chosen to carry out the DInSAR procedures here reported, due to the following reasons:

1. Crop growth monitoring requires spatial resolution of 1 m or less, a growth measurement accuracy of centimeters, short revisit time and an adequate radar wavelength. The drone-borne solution easily fulfills these requirements.
2. Satellite-based DInSAR cannot satisfy all the requirements mentioned above.
3. Aircraft-based DInSAR can meet those conditions; however, the survey costs are not economically feasible for both the research work and the operational case.

This paper is divided as follows. Section 2 presents the drone-borne SAR system; the basics of SAR imaging; a summary of the DInSAR theory; the proposed model for estimating corn crop growth, considering backscattering contribution; a brief description of the experimental site; an outline of the field measurement process; the data acquisition plan for the drone-borne SAR; and the procedures for data processing. Section 3 shows the experimental results, consisting of acquired drone-borne SAR images, and both qualitative and quantitative validations of the attained growth maps. Section 4 discusses those results, and Section 5 presents the authors' conclusions.

2. Materials and Methods

2.1. Drone-Borne SAR System

The radar prototype, described in [12] and shown in Figure 1, operates in three bands—P, L and C—with five channels: two interferometric C-band antennas, two polarimetric L-band antennas and one P-band antenna. This work only shows results for the DInSAR with the L-band HH-polarization, which corresponds to the transmission and reception of signals with horizontal polarization. The two interferometric C-band antennas are used for calculating the digital surface model, DSM, which is then used in the DInSAR calculation. Nominal height accuracy is better than 1 m RMS with a spatial resolution of 2 m. Table 1 shows the main parameters of radar acquisition for the L-band HH-polarization.

Figure 1. Image of the drone-borne SAR with the L-band, the P-band and the two C-band antennas.

Table 1. Main parameters of radar acquisition.

Radar Parameters	Value
Carrier wavelength	22.84 cm
Bandwidth	150 MHz
Polarization	HH
Peak Power	100 mW
Mean Power	1 mW
Pulse Repetition Frequency	10 kHz
Incidence angle	45 deg
Mean drone height	120 m
Mean drone velocity	2 m/s
Maximum acquisition time	20 min
Motion Sensing System, MSS	D-GNSS + IMU
DInSAR accuracy	6 mm
Range resolution	1 m
Azimuth resolution	10 cm
Processed azimuth bandwidth	20 Hz
Processed aperture at 45 deg. incidence angle	196 m
Single-look-complex range sampling	61 cm
Single-look-complex azimuth sampling	5 cm

The Motion Sensing System (MSS), which includes a single channel GNSS receiver and an inertial measurement unit (IMU), is integrated into the radar, as shown in [12]. There is also a ground station with a single channel GNSS, to provide differential GNSS processing.

2.2. SAR Imaging

A circular flight pattern was chosen for generating the growth maps presented in this work. This flight geometry provides images with high resolution, which is given by the following [13]:

$$\delta \approx \frac{1.2c}{\pi \cos(\theta_e) f_c},$$

(1)

where c is the speed of light, f_c is the signal's central frequency and θ_e is the depression angle. Expression (1) is valid for both directions of the ground plane. Additionally, the circular flight pattern helps to reduce shadow effects in the processed image, since it can provide full aspect coverage of the targets of interest [14].

The images were processed by using a time-domain back-projection algorithm, which is a method that is easily applied to nonlinear flight patterns. Let $\vec{r}_l$ be a radar position along the flight track, and let $\vec{p}_{mn}$ be the location of a particular pixel on the image sample grid. The back-projected signal is calculated as follows [15]:

$$s\left(\vec{p}_{mn}\right) = \sum_l g\left(\vec{r}_l, R_{lmn}\right) exp\left(j2k_c R_{lmn}\right),$$

(2)

where k_c is the central wavenumber, and $R_{lmn} = |\vec{r}_l - \vec{p}_{mn}|$.

2.3. DInSAR Theory Description

DInSAR is a type of interferometry that provides information on the terrain height displacement between two flights, at different times, following the same flight path. In the DInSAR case, there is no interferometric baseline, so a topographic map is not possible. However, a terrain height displacement over time is perceptible [16]. In the present case, the analyzed terrain is a crop area.

Figure 2 presents a circular flight pattern over a crop field. The first survey is carried out at a time t_1. The second survey follows the same circular flight pattern of the first flight and occurs at a time t_2. Due to drone instability and weather conditions, the flight patterns are not identical but very similar.

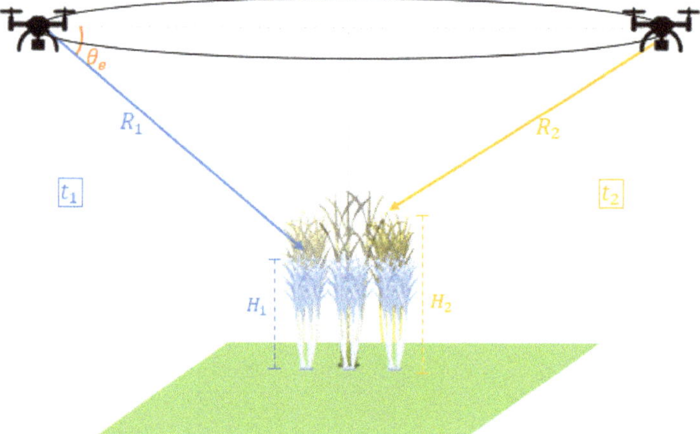

Figure 2. DInSAR circular flight pattern and illumination geometry.

In this example, consider that a first flight is performed on time t_1, during which the crop height is H_1, corresponding to a range R_1 from the radar. Then, on time t_2, another flight is executed, when the crop height is now H_2, and is at a range R_2 from the radar. Let $\Delta R = R_2 - R_1$ be the difference in radar range, and let $\Delta H = H_2 - H_1$ be the growth of the corn crop between these two surveys. It is possible to establish a relationship between these quantities, as shown in Figure 3.

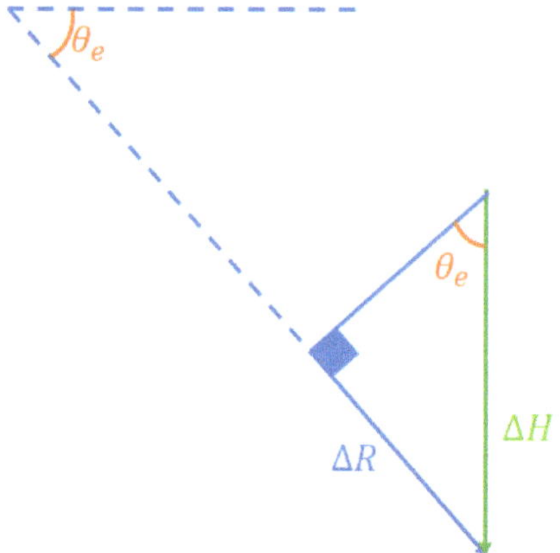

Figure 3. Interaction between radar signal and crop height.

The relationship between height and range is given by the following:

$$\Delta H = \frac{\Delta R}{sin\theta_e},\tag{3}$$

where θ_e is the depression angle. As the drone navigation system performs flight tracks with a position accuracy of about 20 cm at 120 m height, the flight tracks on different dates are assumed to be identical and, therefore, θ_e is virtually the same for any flight. The relationship between the radar range, ΔR, and the phase, $\Delta\varnothing$, can be described as follows [17]:

$$\Delta R = \frac{\lambda}{4\pi}\Delta\varnothing.\tag{4}$$

For every point of interest in the image sample grid, the mean value of θ_e is calculated from the contributions of the depression angles at each drone position along the flight track. If the radar antenna does not illuminate the point of interest at a given drone position, the corresponding contribution is not taken into account. This assessment is based on the antenna aperture angles in azimuth and elevation, considering a circular flight path.

In order to calculate the phase difference between both radar images, it is necessary to determine the phase corresponding to a null vertical displacement or "zero-movement" [16]. That is, it is required to find the phase corresponding to an object within the image for which there was certainly no displacement between the acquisition flights. The zero-movement phase is then subtracted from the phase difference information. Figure 4 shows the scheme used for processing the phase data obtained from two radar images.

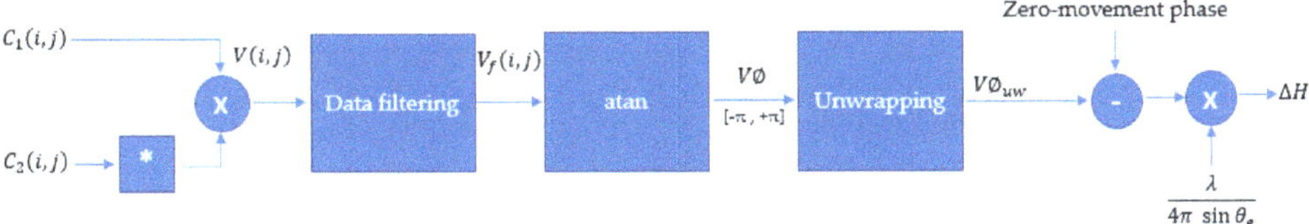

Figure 4. Block diagram for DInSAR processing.

After performing back-projection processing, two images were acquired that have amplitude and phase information. The information of interest is the phase difference between them, calculated as follows:

$$V(i,j) = C_1(i,j) \times C_2^*(i,j), \tag{5}$$

where $C_1(i,j)$ and $C_2(i,j)$ represent the two images, and $V(i,j)$ is the image containing the phase difference information. Then, the real and imaginary components of $V(i,j)$ are filtered, and so the phase information $V\varnothing$ is obtained. The next step is to get $V\varnothing_{uw}$ from $V\varnothing$ by using a technique known as phase unwrapping [18]. After that, the zero-movement phase is subtracted. Finally, the resultant value is multiplied by a factor based on Equations (3) and (4), to obtain the vertical displacement or interferometric height difference, ΔH.

2.4. Estimation Model for Corn Crop Growth

Using the DInSAR model described above to retrieve the height difference from the phase information, and taking into account the reflectivity of the corn crop and the soil with weed plants around it, a model for estimating the growth of a corn crop between two dates is proposed, considering the reflectivity contribution at different development stages of the crop. This model is briefly described in Figure 5.

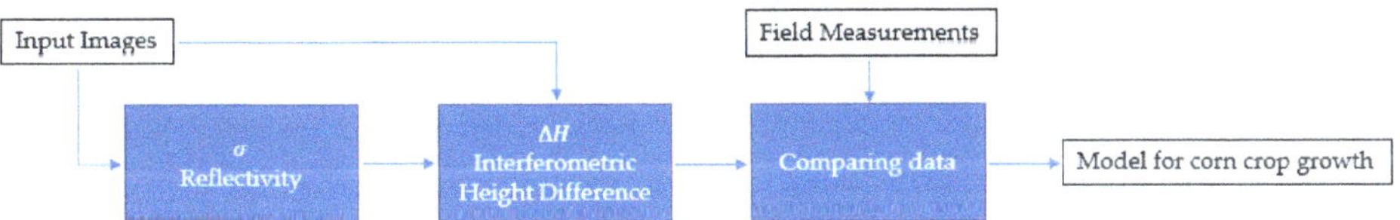

Figure 5. Block diagram for the estimation model for corn crop growth.

The reflectivity data obtained from the input images can be decomposed as follows [2]:

$$\sigma_T = \sigma_{corn} + \sigma_{wp} + \sigma_{db}, \tag{6}$$

where σ_T is the total reflectivity measured by the radar, σ_{corn} represents the corn reflectivity, σ_{wp} is the reflectivity of the soil with weed plants that can be measured in an adjacent area to the corn crop and σ_{db} is the reflectivity of the double-bounce between the crop stem and the soil, as illustrated in Figure 6. At different stages of the crop growth, the proportions of these reflectivities vary. In early stage growth, shown in Figure 6a, the portion due to double bounce is the most significant, while in late-stage growth, pictured in Figure 6b, the contribution of the corn crop is dominant.

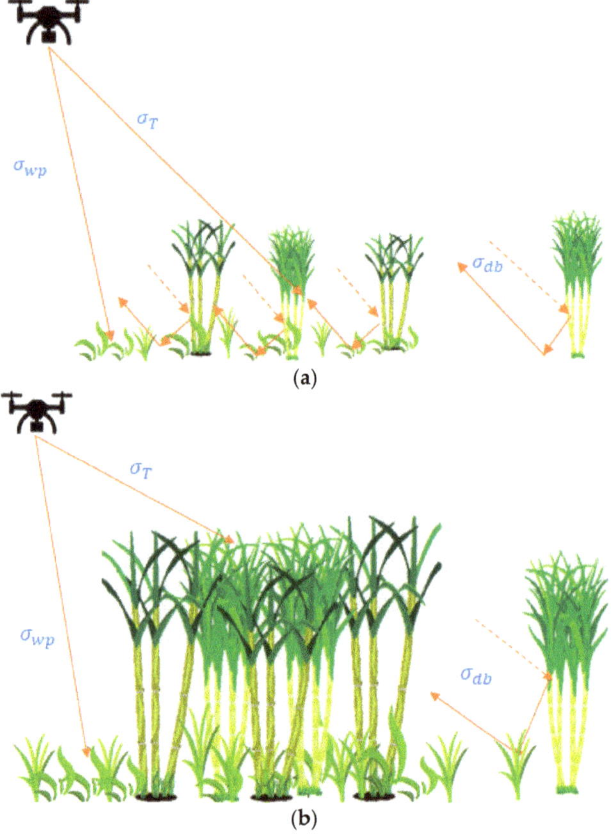

Figure 6. Representation of reflectivity in the corn crop at different growth stages. (**a**) Early stage growth. (**b**) Late-stage growth.

Consider that σ_{wp} can be estimated from the reflectivity of areas close to the crop so that it can be subtracted from Equation (6); thus, the resulting reflectivity, σ_R, is as follows:

$$\sigma_R = \sigma_T - \sigma_{wp} = \sigma_{corn} + \sigma_{db}. \tag{7}$$

Assuming that σ_{corn} is proportional to σ_R by a contribution factor $K_R \in [0, 1]$, then:

$$\sigma_{corn} = K_R \sigma_R. \tag{8}$$

Also, the double-bounce contribution can be estimated from K_R and σ_R as follows:

$$\sigma_{db} = (1 - K_R)\sigma_R. \tag{9}$$

As the presence of stems and branches is dominant during the early stages of crop growth, the contribution factor, K_R, is negligible and, from Equation (9), $\sigma_{db} \approx \sigma_R$. On the other hand, in late stages of crop growth, the double-bounce contribution decreases drastically due to the presence of leaves; thus, the contribution factor $K_R \approx 1$ and, from Equation (8), $\sigma_{corn} \approx \sigma_R$.

To calculate the height difference measured between two different days, consider the two different stages of crop growth shown in Figure 7.

Figure 7. Crop in two different stages. (**a**) Day 1. (**b**) Day 2.

Assume that, on Day 1, the height of the crop is $H_{corn(1)}$, the height of the weed plants is $H_{wp(1)}$ and the difference between them is $H_{(1)} = H_{corn(1)} - H_{wp(1)}$. Similarly, for Day 2, consider that $H_{(2)} = H_{corn(2)} - H_{wp(2)}$. Then, the total estimated height variation, $\Delta H = H_{(2)} - H_{(1)}$, can be calculated as:

$$\Delta H = \Delta H_{corn} - \Delta H_{wp}, \tag{10}$$

where $\Delta H_{corn} = H_{corn(2)} - H_{corn(1)}$ is the height variation corresponding to the crop and $\Delta H_{wp} = H_{wp(2)} - H_{wp(1)}$ is the height variation due to the growth of weed plants. Equation (10) is only valid when contributions can be measured separately.

The radar measures a sum of complex contributions. The amplitude of the return signal is proportional to $\sqrt{\sigma_T}$, and the interferometric phase is proportional to ΔH, as seen in Equations (3) and (4). As the ratio σ_{corn}/σ_T approaches unity, ΔH_{corn} becomes the major contribution in Equation (10). The same reasoning can be made for σ_{wp} and ΔH_{wp}. In that way, a simple solution is to correct Equation (10) by weighting each contribution by its respective reflectivity.

First, let Equation (7) can be rewritten as follows:

$$1 = \frac{\sigma_{corn}}{\sigma_T} + \frac{\sigma_{db}}{\sigma_T} + \frac{\sigma_{wp}}{\sigma_T} = K_{corn}^2 + K_{db}^2 + K_{wp}^2, \tag{11}$$

where

$$K_{corn} = \sqrt{\frac{\sigma_{corn}}{\sigma_T}} = \sqrt{K_R}\sqrt{\frac{\sigma_R}{\sigma_T}}, \tag{12}$$

$$K_{db} = \sqrt{\frac{\sigma_{db}}{\sigma_T}}, \tag{13}$$

$$K_{wp} = \sqrt{\frac{\sigma_{wp}}{\sigma_T}}, \tag{14}$$

are the corn, weed and double-bounce contributing factors, respectively.

Now, assuming that each contribution can be weighted by its respective contributing factor, the difference in height measure by the radar between two consecutive days can be expressed by the following:

$$\Delta H = \Delta H_{corn} K_{corn} - \Delta H_{wp} K_{wp}. \tag{15}$$

Then, from Expressions (12), (14) and (15) comes the following:

$$\Delta H_{corn} = \frac{1}{\sqrt{K_R}} \sqrt{\frac{\sigma_T}{\sigma_R}} \left(\Delta H + \Delta H_{wp} \, K_{wp} \right). \tag{16}$$

Next, a new correction factor $K = 1/\sqrt{K_R}$ is defined based on field measurement data as follows:

$$K = \frac{\Delta H_{corn(fm)}}{\sqrt{\frac{\sigma_T}{\sigma_R} \left(\Delta H + \Delta H_{wp} K_{wp} \right)}}, \tag{17}$$

where $\Delta H_{corn(fm)}$ represents the height difference data measured on the crop. This new correction factor (K) is only valid for the corn model.

Finally, the estimated growth between two dates in a corn crop can be described as follows:

$$\Delta H_{corn-est} = K \sqrt{\frac{\sigma_T}{\sigma_R}} \left(\Delta H_T + H_{wp} K_{wp} \right). \tag{18}$$

2.5. Experimental Site

The experimental site covers an area of 300 m × 300 m, located at the School of Agricultural Engineering of the University of Campinas (UNICAMP), as shown in Figure 8. The survey flights occurred on the following dates: 11 December 2018; and 17 April, 2 July, 17 July and 22 August 2019.

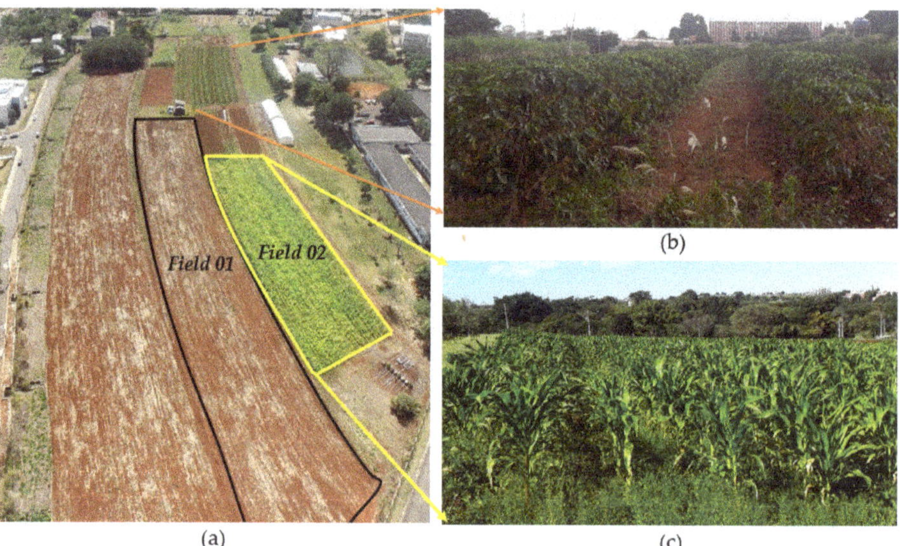

Figure 8. Overview of the experimental site at UNICAMP. (**a**) Aerial View with fields 01 and 02 at the bottom and the coffee crop at the top, on 22 August 2019. Field 01 was bare. (**b**) Ground photo of the coffee crop. (**c**) Ground photo of the corn crop on field 02.

2.6. Field Measurements

Height measurements of coffee and corn crops, depicted in Figure 8b,c, respectively, were carried out over nine months. In this period, the corn reached the maturity stage in December 2018, and a new crop was set in July 2019. Coffee reached the maturity stage in March 2019. The corn crop is structured with 14 rows, where 20 height measurements were made per row. The heights were measured from the soil to the top of the corn crop. The following measurements were taken on the same days of the flight surveys: 2 July, 17 July and 22 August 2019. The field measurements of the corn crop height are subjected to the natural variability of visual measurement errors.

During the survey period, corn was sown in two areas: field 01 and 02, as shown in Figure 8a. Hereafter, fields 01 and 02 will be labeled as C1 and C2, respectively, where "C" denotes the corn culture.

2.7. Drone-Borne SAR Data Acquisition

Three trihedral corner reflectors with square sides and an edge length of 0.6 m were used as a ground and radiometric calibration reference. The resulting radar cross-section is 20 dBsm.

On each acquisition date, the experiment with the drone-borne system was executed as follows:

- Mount three corner reflectors on the test site, for planimetric and radiometric calibration purposes;
- Place the GNSS ground station close to the initial position of the drone and start the GNSS recording;
- Perform each flight over the experimental site, following the subsequent procedure: turn on the drone and the radar, wait 15 min for simultaneous and stationary recording of ground station and radar GNSS data, take-off, perform the same circular flight track, land, wait 15 min for simultaneous and stationary recording of ground station and radar GNSS data, and turn-off the radar and the drone;
- Dismount the GNSS ground station and the drone. Download the acquired data for processing.

2.8. Drone-Borne SAR Data Processing

The collected data were processed as follows:

- Differential GNSS processing of the ground station and the radar GNSS receivers;
- IMU and differential GNSS data fusion for generating position and antenna orientation history;
- Radar data processing at each acquisition date, according to Section 2.2: range compression and back-projection for the azimuth compression. The output is a geocoded single-look-complex (SLC) image;
- Verification of the absolute position of the corner reflectors in the geocoded SLC images;
- Differential interferometric processing with data from previous acquisitions, as defined in Section 2.3;
- Production of the crop growth map, as described in Section 2.4;
- Generation of the corresponding multi-look images with 30 cm × 30 cm sampling.

3. Results

3.1. Drone-Borne SAR Images

The circular flights over the experimental site were carried out by following the path shown in Figure 9. Figure 10 displays the SAR images acquired with circular flights on different acquisition dates. Each image depicts an area of 300 m × 300 m, with 30 cm sampling in both directions.

Figure 10b shows the area C2 with the corn crop in a mature stage. The new seeding of area C2 took place in July 2019, shown in Figure 10d, and this area presents increasing reflectivity through Figure 10e,f, until reaching its maturity.

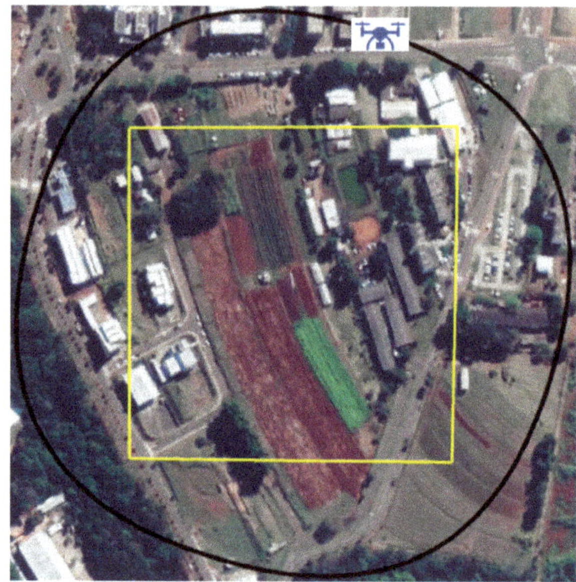

Figure 9. Description of the circular flight. In black, the circular flight track, and in yellow, the study area of 300 m × 300 m.

3.2. Qualitative Validation of Growth Deficit Maps

DInSAR technique was applied to the campaigns of 5 December and 11 December 2018, and 17 April 2019, on coffee, sugar cane and corn crops. According to the field data, the corn crop was present only on the first two dates mentioned, and the other crops had an insignificant height variation between the three campaign dates. DInSAR provided a compatible result for these crops, showing a negligible height variation during those dates.

The first trial for a crop growth map was performed with the coffee crop, presented in Figure 8. No field measurement of the crop height variation was determined. However, a rough growth estimate of about 10 cm was expected, as the coffee crop was quite mature in December. Because of a small morphology alteration in the crop due to its maturity, the interferogram had a low but acceptable coherence of 0.2.

A growth map corresponding to a coffee crop of approximately 1700 m^2 is shown in Figure 11. The mean height growth between those dates with DInSAR is 11 cm, with a standard deviation of 6 cm. The south area of the coffee crop presented less growth than the north area. A mask was used to discard all targets that do not correspond to the coffee crop. Moreover, a 15 × 15 moving average filter was used over the interferogram, with a pixel spacing of 30 cm × 30 cm. Only data with a coherence greater than 0.1 were considered valid data. This result has motivated a more in-depth study of DInSAR for growth maps.

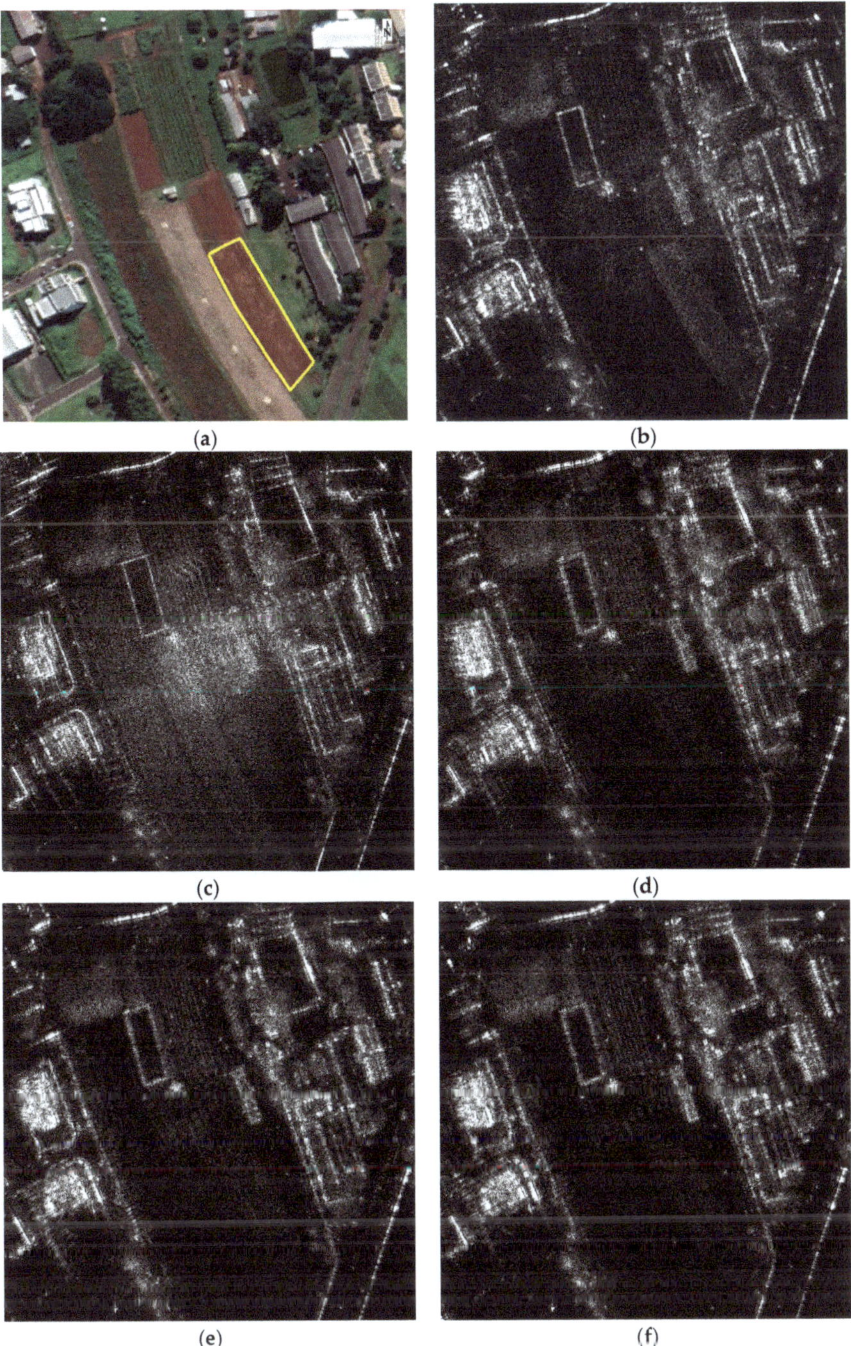

Figure 10. Temporal comparison between (**a**) optical image from Google Earth and drone-borne SAR images of the circular flight tracks, acquired on (**b**) 11 December 2018; (**c**) 17 April 2019; (**d**) 2 July 2019; (**e**) 17 July 2019; and (**f**) 22 August 2019. Field C2 has a yellow border in (**a**).

Figure 11. Growth map of the coffee crop from 11 December 2018, to 17 April 2019.

Figure 12 shows the corresponding SAR images for the acquisition dates of 11 December 2018, and 17 April 2019, where the yellow squares represent the coffee crop for which the growth map of Figure 11 was produced.

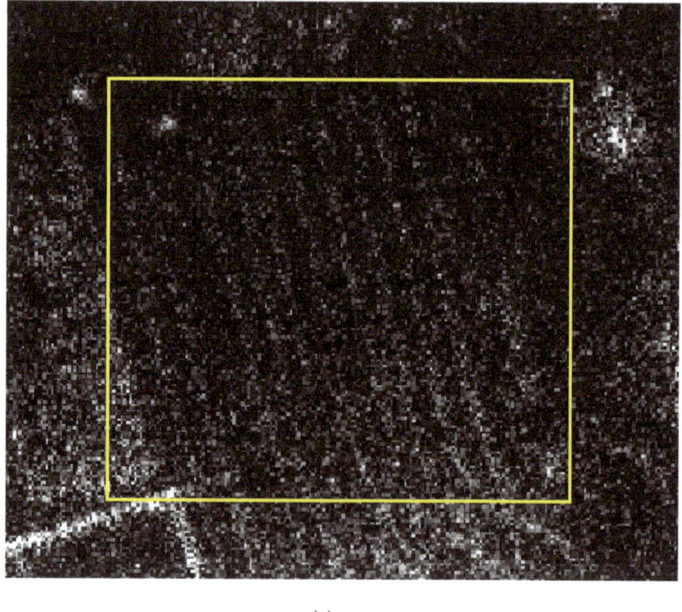

(a)

Figure 12. *Cont.*

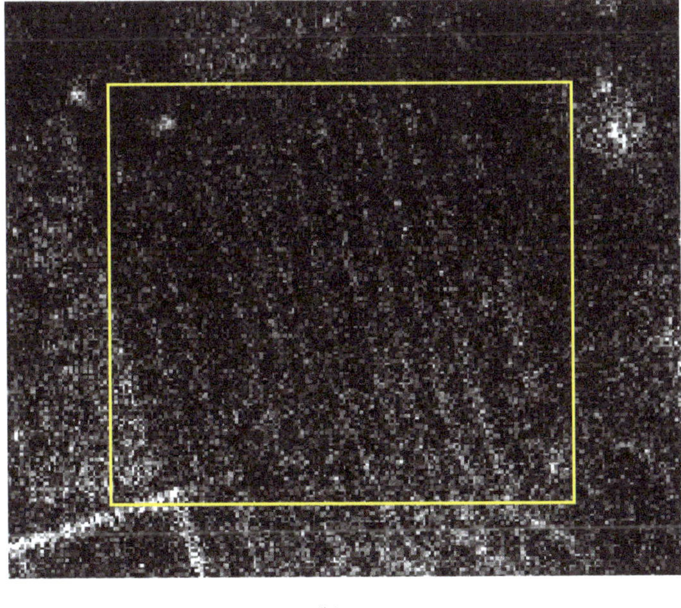

(b)

Figure 12. SAR images of coffee crop acquired on (**a**) 11 December 2018; and (**b**) 17 April 2019.

3.3. Quantitative Validation of Growth Deficit Maps

SLC images of circular flight tracks were used to validate the growth estimation model considering the reflectivity contribution. The high resolution allows a much easier analysis, as the cornfield C2 is rather small.

To obtain the phase reference, known as the zero-movement phase, an object that would remain static during all flights was searched. Initially, it was thought to use the dihedral corner reflectors. However, it would not be possible to leave the dihedral corner reflectors at all times in the study area, because other daily activities are carried out there. Besides, a dihedral corner reflector can only be observed during a small portion of the circular track. Therefore, a metal fence near the crop area was chosen as a reference, as it is a fixed object and can be observed during the entire flight (see Figure 13).

(a)

Figure 13. *Cont.*

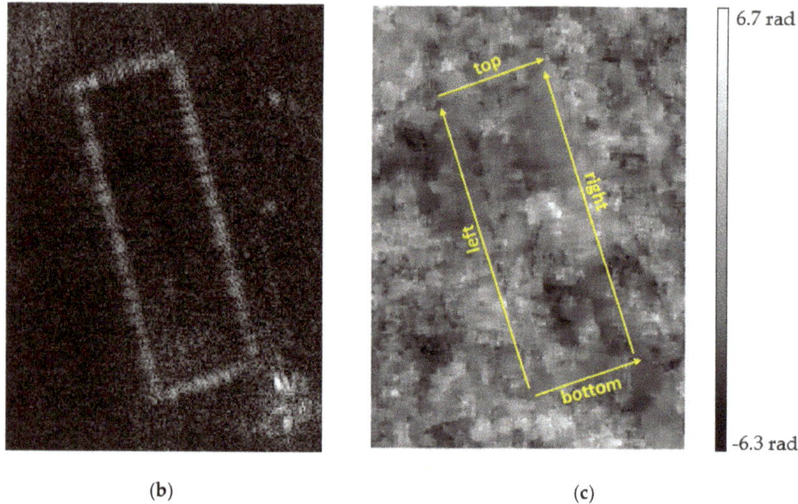

(b)   (c)

Figure 13. (**a**) Optical image of the metal fence; (**b**) radar image of the metal fence; and (**c**) unwrapped interferogram corresponding to the metal fence.

To obtain the zero-movement phase, profiles were drawn for the interferogram over the area corresponding to the metal fence, Figure 13c. The profiles are shown in Figure 14a,b. The zero-movement phase was calculated by using the average of the profiles, which, in this example, resulted in 0.76 rad.

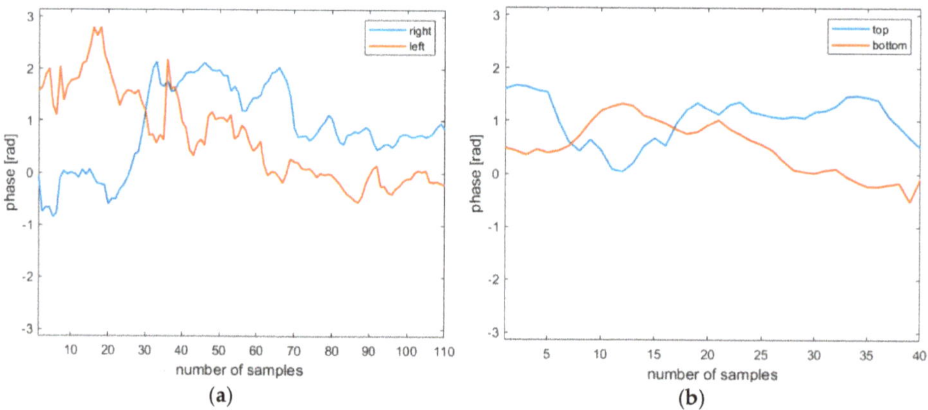

Figure 14. (**a**) Right and left side profiles of the fence phase; (**b**) top and bottom side profiles of the fence phase.

Then, the height growth of the cornfield C2 was monitored, as shown in Table 2. These data, together with the reflectivity data for the corn crop and a neighboring area (see Figure 6), were used to estimate the correction factor, K, of Equation (17), which varies along with the phenological life of the corn crop.

Table 2. Corn crop field measurements.

Height Measurement Date	Days after Planting	Corn Height
2 July 2019	48	68 cm
17 July 2019	67	99 cm
22 August 2019	99	125 cm

Figure 15 displays three values for the correction factor, K, estimated by using the available radar data and field measurement data from the three different periods presented in Table 2. Figure 15 also shows the fitted curve (blue line) for the correction factor, K, as a function of the corn phenology stage.

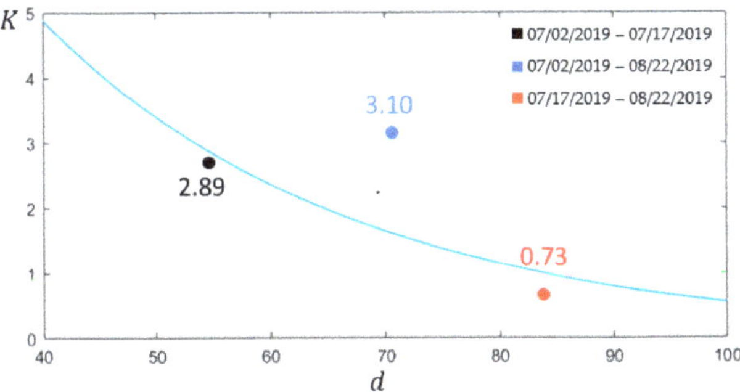

Figure 15. Correction factor (K) as a function of the corn phenology stage.

Due to the small amount of data, there is some flexibility when choosing the type of equation that best fits the experimental data. An exponential fitting curve was chosen, and the resultant expression is as follows:

$$K = 20.75 \exp(-0.03617d), \tag{19}$$

where d represents the average number of days after planting, calculated over each data collection period, as displayed in Table 3. Equation (19) is a first attempt to find an expression for K, using the few available data, and it is calculated with MATLAB Curve Fitting Toolbox [19].

Table 3 also shows the height-difference data between two dates obtained by field measurements in the corn crop and estimates from the proposed model, using the radar data together with the correction factor, K.

Table 3. Comparison between field measurement data and estimated radar data in the corn crop for height difference information.

Data Collection Period	d	Height Difference (Field Measurement Data)	Height Difference (Estimated Radar Data)
2 July 2019–17 July 2019	57 days	31 cm	36 cm
2 July 2019–22 August 2019	73 days	57 cm	42 cm
17 July 2019–22 August 2019	83 days	26 cm	28 cm

Figure 16 presents the crop growth maps derived from the same acquisition intervals displayed in Figure 15. Rapid growth can be seen from 2 July to 17 July, though growth is noticeably slower between 17 July and 22 August.

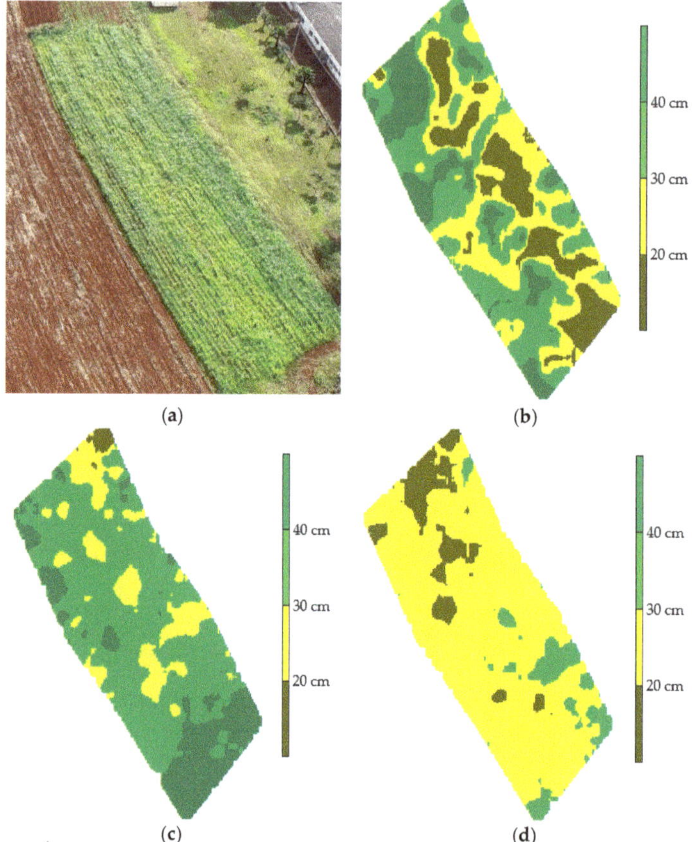

Figure 16. Crop growth maps derived from three acquisition intervals: (**a**) optical reference image from corn crop (**b**) from 2 July to 17 July 2019; (**c**) from 2 July to 22 August 2019; and (**d**) from 17 July to 22 August 2019.

4. Discussion

Determining the zero-movement phase is an essential step for estimating crop growth. The zero-movement phase is not null due to external factors such as the system's thermal noise, the difference between flight paths from distinct dates and internal changes of the radar system between different flights [16]. The chosen reference object was a metal fence because it is always fixed, thus providing more robust results than moving objects, for which precise positioning between separate flights is not guaranteed. Moreover, it can be seen from all directions in a circular flight track, and it is typically found near crop fields.

Corner reflectors are not practical reference points for DInSAR, as they should be well fixed and kept clean during the DInSAR observation period. The authors intend to continue using corner reflectors, only for purposes of planimetric accuracy and height calibration. Luneburg lenses could also be used to better calibrate the back-projection processor [20].

No other publications were found that use the DInSAR technique on a corn crop to estimate growth, but some similar works are noteworthy. Erten et al. [10] used the DInSAR technique, with the TANDEM-X satellite, over a rice cultivation area, and were able to estimate growth with an RMSE of 18 cm. Cao et al. [21] used an airborne L-band radar system to measure landslide with a precision of a few centimeters.

In this work, a first attempt to monitor growth was carried out with a coffee crop, as described in Section 3.2. The coffee had a barely noticeable height variation between 11 December 2018, and 17 April 2019. However, using DInSAR, it was possible to measure an 11 cm growth, with a standard deviation of 6 cm. No model will estimate the growth of a crop that hardly grows.

As for the corn crop growth using correction factor K, the errors obtained from 2 July to 17 July 2019, and from 17 July to 22 August 2019, were 5 cm and 2 cm, respectively, while from 2 July to 22 August 2019, the error was 15 cm. Therefore, the longest period between data acquisitions produces the largest error, whereas shorter acquisition periods provide more accurate results. Therefore, to achieve a more accurate growth model oriented to the corn crop, it is necessary to retrieve a better estimation of the variable K, and thus more flights are required. Another way to develop this model would be to use the two interferometric C-band antennas to estimate crop height [22].

The corn crop was chosen due to its rapid growth. Nevertheless, this work could be extended to other crops, such as sugar cane or coffee. The case of sugar cane, which grows slower than corn, is of particular interest to the authors; therefore, more campaigns will be needed over time.

5. Conclusions

This work proposes a novel method for estimating the growth of different crops by executing circular flight paths with a drone-borne DInSAR operating in the L-band with HH polarization. First tests on late-stage coffee, corn and sugar cane crops have shown that it is possible to reliably estimate small height variations. In the case of coffee, the growth within approximately four months was estimated at 11 cm, with a standard deviation of 6 cm. Such a growth rate is difficult to perceive visually or to measure with conventional tools.

Furthermore, a method is proposed to estimate the growth of a corn crop, taking into account its phenological development. Although more campaigns on the area of interest are still necessary, these first-stage results show a strong agreement with the measured field data. The largest error was 15 cm, corresponding to the longest period between data acquisitions, for which the crop growth was approximately 55 cm. In contrast, errors of up to 5 cm were obtained for shorter acquisition periods.

The images obtained from the circular flights were processed with the back-projection algorithm using a 30 cm × 30 cm sampling. In future work, a more accurate growth estimation is expected since back-projection images shall be processed with a 5 cm × 5 cm sampling.

The authors are motivated to continue with this line of research, taking as a starting point the methods and first-stage results obtained so far. More consistent results are expected in the near future, when more data will be available.

Author Contributions: Conceptualization, G.O. and H.E.H.-F.; investigation, methodology, software and validation, G.O., M.S.A., J.A.G., V.C., L.S.B. and F.C.; validation, J.Y. and D.L.; formal analysis and writing, G.O., M.S.A., J.A.G., L.P.O., V.C., F.C. and L.F.M.; supervision, review and funding acquisition, L.P.O., B.T., D.L., L.H.G. and H.E.H.-F. All authors have read and agreed to the published version of the manuscript.

Funding: This research was funded by government agencies CAPES and CNPq and the São Paulo State agency FAPESP, under the contracts PITE 2017/19416-3 and PIPE 2018/00601-8.

Acknowledgments: We thank the teams of the schools of Electrical and Computer Engineering and of Agricultural Engineering of the University of Campinas, UNICAMP, for the given support.

References

1. Liu, C.-A.; Chen, Z.-X.; Shao, Y.; Chen, J.-S.; Hasi, T.; Pan, H.-Z. Research advances of SAR remote sensing for agriculture applications: A review. *J. Integr. Agric.* **2019**, *18*, 506–525. [CrossRef]
2. Lin, H.; Chen, J.; Pei, Z.; Zhang, S.; Hu, X. Monitoring Sugarcane Growth Using ENVISAT ASAR Data. *IEEE Trans. Geosci. Remote Sens.* **2009**, *47*, 2572–2580. [CrossRef]
3. Baghdadi, N.; Cresson, R.; Todoroff, P.; Moinet, S. Multitemporal Observations of Sugarcane by TerraSAR-X images. *Sensors* **2010**, *10*, 8899–8919. [CrossRef] [PubMed]

4. Skriver, H. Crop Classification by Multitemporal C- and L-Band Single- and Dual-Polarization and Fully Polarimetric SAR. *IEEE Trans. Geosci. Remote Sens.* **2012**, *50*, 2138–2149. [CrossRef]

5. Liu, X.; Jiao, L.; Tang, X.; Sun, Q.; Zhang, D. Polarimetric Convolutional Network for PolSAR Image Classification. *IEEE Trans. Geosci. Remote Sens.* **2019**, *57*, 3040–3054. [CrossRef]

6. Lin, K.-F.; Perissin, D. Single-Polarized SAR Classification Based on a Multi-Temporal Image Stack. *Remote Sens.* **2018**, *10*, 1087. [CrossRef]

7. Intarat, T.; Rakwatin, P.; Srestasathien, P.; Triwong, P.; Tangsiriworakul, C.; Noivum, S. Potential of Sugar cane monitoring using Synthetic Aperture Radar in Central Thailand. In Proceedings of the 36th Asian Conference of Remote Sensing, Quezon City, Philippines, 19–23 October 2015.

8. Rossi, C.; Erten, E. Paddy Rice Monitoring Using TanDEM-X. *IEEE Trans. Geosci. Remote Sens.* **2015**, *53*, 900–910. [CrossRef]

9. Wang, H.; Magagi, R.; Goita, K. Polarimetric Decomposition for Monitoring Crop Growth Status. *IEEE Geosci. Remote Sens. Lett.* **2016**, *13*, 870–874. [CrossRef]

10. Erten, E.; Lopez-Sanchez, J.; Yuzugullu, O.; Hajnsek, I. Retrieval of agricultural crop height from space: A comparison of SAR techniques. *Remote Sens. Environ.* **2016**, *187*, 130–144. [CrossRef]

11. Frey, O.; Werner, C.L.; Coscione, R. Car-borne and UAV-borne mobile mapping of surface displacements with a compact repeat-pass interferometric SAR system at L-band. In Proceedings of the IGARSS 2019—2019 IEEE International Geoscience and Remote Sensing Symposium, Yokohama, Japan, 28 July–2 August 2019; pp. 274–277.

12. Moreira, L.; Castro, F.; Góes, J.A.; Bins, L.; Teruel, B.; Fracarolli, J.; Castro, V.; Alcântara, M.; Oré, G.; Luebeck, D.; et al. A Drone-borne Multiband DInSAR and Applications. In Proceedings of the 2019 IEEE Radar Conference (RadarConf), Boston, MA, USA, 22–26 April 2019; pp. 1–6.

13. Ishimaru, A.; Chan, T.-K.; Kuga, Y. An imaging technique using confocal circular synthetic aperture radar. *IEEE Trans. Geosci. Remote Sens.* **1998**, *36*, 1524–1530. [CrossRef]

14. Palm, S.; Sommer, R.; Pohl, N.; Stilla, U. *Airborne SAR on Circular Trajectories to Reduce Layover and Shadow Effects of Urban Scenes*; International Society for Optics and Photonics: Edinburgh, UK, 2016; p. 100080N.

15. Doerry, A.W.; Bishop, E.E.; Miller, J.A. *Basics of Backprojection Algorithm for Processing Synthetic Aperture Radar Images*; Sandia National Laboratories: Albuquerque, NM, USA, 2016; pp. 1–57.

16. Richards, J.A. *Remote Sensing with Imaging Radar*; Signals and Communication Technology; Springer: Heidelberg, Germany, 2009; ISBN 9783642020193.

17. Hanssen, R.F. *Radar Interferometry Data Interpretation and Error Analysis*; Remote Sensing and Digital Image Processing; Kluwer Academic: Dordrecht, The Netherlands, 2001; ISBN 9780792369455.

18. Giglia, D.C.; Romero, L.A. Robust two-dimensional weighted and unweighted phase unwrapping that uses fast transforms and iterative methods. *J. Opt. Soc. Am.* **1994**, *11*, 107–117. [CrossRef]

19. *Curve Fitting Toolbox*; The Math Works: Natick, MA, USA, 2018.

20. Ponce, O.; Prats-Iraola, P.; Scheiber, R.; Reigber, A.; Moreira, A. First Airborne Demonstration of Holographic SAR Tomography with Fully Polarimetric Multicircular Acquisitions at L-Band. *IEEE Trans. Geosci. Remote Sens.* **2016**, *54*, 6170–6196. [CrossRef]

21. Cao, N.; Lee, H.; Zaugg, E.; Shrestha, R.; Carter, W.; Glennie, C.; Wang, G.; Lu, Z.; Fernandez-Diaz, J.C. Airborne DInSAR Results Using Time-Domain Backprojection Algorithm: A Case Study Over the Slumgullion Landslide in Colorado With Validation Using Spaceborne SAR, Airborne LiDAR, and Ground-Based Observations. *IEEE J. Sel. Top. Appl. Earth Obs. Remote Sens.* **2017**, *10*, 4987–5000. [CrossRef]

22. Duerch, M. Backprojection for Synthetic Aperture Radar. Ph.D. Thesis, Brigham Young University, Provo, UT, USA, 2013.

 *remote sensing*

Article

Small Multicopter-UAV-Based Radar Imaging: Performance Assessment for a Single Flight Track

Ilaria Catapano [1,*], Gianluca Gennarelli [1], Giovanni Ludeno [1], Carlo Noviello [1], Giuseppe Esposito [1,2], Alfredo Renga [2], Giancarmine Fasano [2] and Francesco Soldovieri [1]

[1]   Institute for Electromagnetic Sensing of Environment (IREA), National Research Council (CNR), 80124 Napoli, Italy; gennarelli.g@irea.cnr.it (G.G.); ludeno.g@irea.cnr.it (G.L.); noviello.c@irea.cnr.it (C.N.); esposito.g@irea.cnr.it (G.E.); soldovieri.f@irea.cnr.it (F.S.)

[2]   Department of Industrial Engineering (DII) University of Naples "Federico II" via Claudio 21, 80124, Naples, Italy; alfredo.renga@unina.it (A.R.); g.fasano@unina.it (G.F.)

*   Correspondence: catapano.i@irea.cnr.it; Tel.: +39-0817620656

Received: 18 December 2019; Accepted: 27 February 2020; Published: 29 February 2020

Abstract: This paper deals with a feasibility study assessing the reconstruction capabilities of a small Multicopter-Unmanned Aerial Vehicle (M-UAV) based radar system, whose flight positions are determined by using the Carrier-Phase Differential GPS (CDGPS) technique. The paper describes the overall radar imaging system in terms of both hardware devices and data processing strategy for the case of a single flight track. The data processing is cast as the solution of an inverse scattering problem and is able to provide focused images of on surface targets. In particular, the reconstruction is approached through the adjoint of the functional operator linking the unknown contrast function to the scattered field data, which is computed by taking into account the actual flight positions provided by the CDGPS technique. For this inverse problem, we provide an analysis of the reconstruction capabilities by showing the effect of the radar parameters, the flight altitude and the spatial offset between target and flight path on the resolution limits. A measurement campaign is carried out to demonstrate the imaging capabilities in controlled conditions. Experimental results referred to two surveys performed on the same scene but at two different UAV altitudes verify the consistency of these results with the theoretical resolution analysis.

Keywords: radar imaging; unmanned aerial vehicle; inverse scattering; linear scattering models; global positioning systems

---

1. Introduction

Radar imaging performed by UAV platforms [1], and more in detail by M-UAV platforms [2], is attracting huge attention in remote sensing community as a cost-effective solution to cover wide and/or not easily accessible regions, with high operative flexibility [3]. Indeed, M-UAVs have vertical lift capability, allow take-off and landing from very small areas without the need for long runways or dedicated launch and recovery systems, and are able to move in all directions. These peculiar features allow their use at any location [3] and under different flight modes, thus introducing new possibilities in radar imaging measurements [4]. For instance, M-UAV vertical lift capability can be exploited to perform vertical apertures and implement high-resolution vertical tomography, which is useful in structural monitoring [5]. On the other hand, circular flights are suitable to generate holographic and tomographic radar images [6]. Furthermore, M-UAVs allow waypoint flights in autopilot mode and pre-programmed flights with auto-triggering. This introduces the possibility of designing sophisticated flight strategies, such as specific grid acquisitions devoted to investigating the area of interest or repeat-pass tracks aimed at performing interferometric acquisitions [7].

UAV-based radar imaging receives huge attention in several military and civilian applications, such as surveillance, security, diagnostics, monitoring in civil engineering, cultural heritage and earth observation, with particular emphasis on natural disasters, which should be safely and timely monitored [8]. At the state-of-the-art, radar imaging performed by M-UAVs has been proposed for precision farming [9], forest mapping [10] and glaciology [11]. In addition, M-UAVs have been exploited to perform Synthetic Aperture Radar (SAR), avoiding large platforms when monitoring small areas. In this frame, the first experimentation concerning interferometric P and X band SAR systems onboard UAV platforms has been reported in [12], while a UAV polarimetric SAR imaging system has been proposed in [13]. UAVs have been also exploited in the field of landmine detection as platforms equipped with Ground Penetrating Radar (GPR) systems [14,15].

Despite these promising examples, the development of radar systems onboard M-UAV is at an early stage and M-UAV radar imaging still represents a scientific challenge, especially when small and light M-UAV platforms are deployed. Indeed, the full exploitation of smart and flexible M-UAV imaging radar systems requires the development of reconstruction approaches able to deal with non-conventional data acquisition configurations, where data are not collected along a straight linear trajectory due to environmental conditions or presence of obstacles. In this respect, it is worth pointing out that radar imaging, i.e., the possibility to obtain a focused image of the investigated region, strongly depends on the accurate knowledge of platform position and orientation during the flight. Indeed, platform-positioning errors, comparable with the wavelength of the electromagnetic signal emitted by the radar and not properly compensated, distort the resulting image severely [16]. Moreover, the capability of small M-UAVs to follow straight linear trajectories is limited. Therefore, in order to avoid detrimental effects on radar imaging performance, the compensation of three-dimensional (3D) deviations, with respect to the ideal flight track, represents a key issue requiring accurate knowledge of the UAV platform position and velocity. However, the quality of such information strongly depends on the accuracy of both the embarked navigation sensors and the deployed ground-based tracking devices.

The most popular navigation sensors include the Inertial Measurement Unit (IMU) with gyroscopes and accelerometers, magnetometers and Global Navigation Satellite System (GNSS) receivers (GPS and multi-constellation receivers). These information sources are typically fused together, with the final positioning accuracy that is driven by GNSS and can reach centimeter-level in carrier-phase differential architectures. An additional constraint for small M-UAVs is related to the maximum payload mass that can be embarked limiting the number and typology of onboard electronic devices. In addition, as pointed out in [17], the standalone GPS technology does not allow the positioning accuracy required for high-resolution radar imaging, especially as far as the platform height is concerned. To face this issue, the authors proposed in [17] an edge detection procedure for estimating the flight height with centimeter-level accuracy from radar data, by making the assumption that the terrain is flat and that extended objects on the ground are absent. An alternative approach is the use of ground-based tracking devices, e.g., radars and laser scanners led by a Robotic Total Station (RTS) [18]. These ground-based systems allow trajectory accuracy at the millimeter scale, but only in fully controlled lab-like conditions or in the presence of multiple tracking devices located all around the operative area. A further approach for the correction of the trajectory errors has been proposed in [19] based on SAR imaging autofocus. The SAR imaging autofocusing technique performs the 3D GPS positioning correction by minimizing the Shannon entropy of the scattering function of the volumetric scene under test, where the targets are imaged, by using a back-projection approach [19]. However, the performance of the autofocusing algorithm is affected by the presence of noise and clutter disturbance, which degrade the autofocusing capability.

As a contribution to this issue, in this paper, we present an improved version of the M-UAV radar system in [17] where the accuracy about the position and velocity of M-UAV is enabled by the use of the (single-frequency) Carrier-Phase Differential GPS (CDGPS) technique [20]. This is made possible with the use of an additional ground-based GPS receiver, which allows the offline implementation of the CDGPS technique by exploiting the RTKlib software [21]. Differently from the exploitation of an

RTS, the CDGPS technique can achieve centimeter accuracy even in harsh operation scenarios where an unobstructed line of sight (LOS) between the ground-based GPS receivers and the flying platform may occur [20].

Once an accurate estimate of the 3D flight path has been obtained, a high-resolution radar imaging algorithm is proposed for the case of a single flight track. The radar imaging approach is able to account for the spatial coordinates of the measurement points provided by the CDGPS technique and states the radar imaging as an electromagnetic inverse scattering problem. The inverse problem is linearized by resorting to the Born approximation [22] and the inversion is carried out by means of the adjoint operator [23]. The reconstruction capabilities of the proposed radar imaging system are investigated in terms of resolution limits by a theoretical/numerical analysis, which makes it possible to foresee how the measurement parameters (location of the measurement points and working frequency band) affect the reconstruction performance [24]. Finally, a measurement campaign carried out at an authorized site for amateur UAV testing flights in Acerra, a small town on the outskirts of Naples (Italy), is presented as an experimental assessment of the integrated use of the CDGPS positioning procedure and the adopted imaging approach. The experimental results provide a proof of concept of the imaging performance of the proposed small M-UAV-based radar imaging system.

The paper is organized as follows. Section 2 describes the small M-UAV-based radar imaging system and the strategy adopted to estimate the actual flight path. Section 3 deals with data processing and presents the reconstruction performance analysis. Section 4 reports the experimental validation of the small M-UAV-based radar imaging system. A final discussion on the system performance and the achieved results is reported in Section 5 and conclusions end the paper in Section 6.

2. Imaging System

The small M-UAV imaging system already presented in [17] is improved with a second ground-based GPS station in order to exploit the CDGPS technique (see Figure 1). The system has the following main components, which are briefly described (see [17] for more details):

- *Small M-UAV platform*: DJI F550 hexacopter able to fly at very low speeds (about 1 m/s), thus ensuring a small spatial sampling step and the ability to take-off and land from a very small area;
- *Radar system*: Pulson P440 radar is a light and compact time-domain device transmitting ultra-wideband pulses (about 1.7 GHz bandwidth centered at the carrier frequency of 3.95 GHz) with a low power consumption [25]. The radar system is mounted rigidly on the UAV body (strapdown installation) and no gimbal is adopted. The limited altitude dynamics experienced during flights (very low ground speed and wind speed conditions resulting in small and almost constant roll/pitch angles), the relatively large radar antenna lobes and the limited baseline between the radar antenna and the drone center of mass are such that altitude/pointing knowledge does not play a significant role;
- *GPS receivers/antennas*: two single-frequency Ublox LEA-6T devices are chosen, one mounted onboard the UAV and the other one used as a ground-based station. Both are connected to an active patch antenna. The antenna is directly placed on the ground (Figure 1b) in order to get from CDGPS a direct estimate of the height above ground for the antenna mounted on the drone;
- *CPU controller*: Linux-based Odroid XU4 is devoted to managing the data acquisition for both radar system and onboard GPS receiver, while assuring their time synchronization.

The possibility to estimate the trajectory of the UAV platform depends on the quality of the embarked navigation sensors. By using a standalone onboard GPS receiver, the achievable absolute positioning accuracy is given into a global reference frame, such as WGS84 (World Geodetic System 1984), and is defined according to the specifications provided by the US Department of Defense [26]. Absolute GPS localization errors are estimated as the product of the User Equivalent Range Error (UERE), which is the effective accuracy of the localization errors along the pseudo-range direction, the Horizontal Dilution of Precision (HDOP) and Vertical Dilution of Precision (VDOP). These latter are dimensionless quantities expressing the effect of satellites-receiver geometry. Representative values of UERE, HDOP and VDOP, for good GPS visibility conditions, are, respectively, 3.5 m and 6.6 m [27].

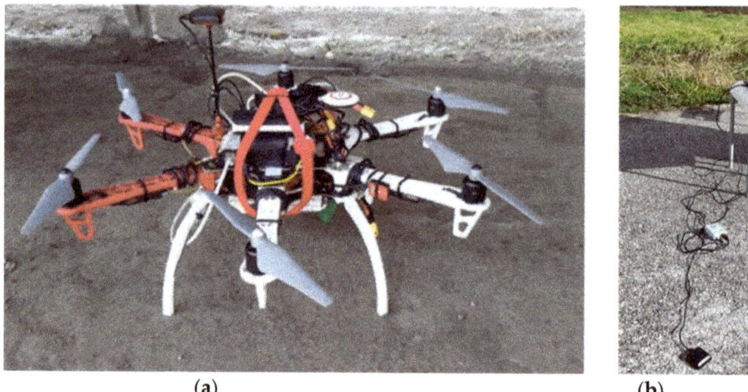

(a)   (b)

Figure 1. M-UAV radar imaging system: (**a**) hexacopter with onboard equipment; (**b**) ground-based GPS station.

When reasonably short time flights are considered, several error sources (i.e., broadcast clock, broadcast ephemeris, group delay, ionospheric delay and tropospheric delay) are strongly correlated both in space and time [20] and introduce a positioning error, which is an almost constant but unknown bias. In addition, the use of a proper processing strategy, such as carrier-smoothing [20], allows a reduction of the measurement noise [28], thus improving the standalone GPS performance.

As shown in [17], in the frame of radar imaging, it is important to have accurate knowledge of the relative positions of the UAV radar system with respect to the investigated spatial region. Therefore, the constant and unknown bias affecting the horizontal positions provided by a standalone GPS does not play any role in focusing the targets (which, however, will not be reliably localized in the WGS84 reference system), whereas the bias occurring into the vertical position (UAV height) may prejudice satisfactory radar imaging performance.

In this paper, we exploit a strategy based on the use of a CDGPS, which is a method for improving the positioning or timing performance of GPS by exploiting at least one motionless GPS receiver working as a reference station. Here, the CDGPS method is implemented by using two GPS receivers (one mounted onboard the UAV and the other one used at reference ground station), which store the data into a local hard drive.

Each receiver collects single-frequency observables, which is a pseudo-range and a carrier-phase measure for any tracked GPS satellite. It is well-known that carrier-phase measures show significantly reduced measurement noise (in the order of 1/100 of signal wavelength, i.e., mm scale) with respect to pseudo-range ones, but ambiguities appear, so carrier-phase are biased measurements [28,29]. If one is able to resolve the ambiguity, very high accuracy positioning is enabled. This can be achieved by differential techniques, i.e., CDGPS, where differences between the measurements collected by two relatively close receivers are computed. Such differential measures are not affected by common errors between the receivers, due to ionosphere, troposphere and clock errors, so suitable processing is implemented to filter out pseudo-range noise thus deriving an estimate of carrier-phase ambiguities. If a connection through a radio link is established between the UAV and the ground station, CDGPS processing can be performed in real-time, which is defined as "Real-Time Kinematic" (RTK). Offline CDGPS processing is, instead, used in this work, which is typically referred to as "Post-Processing Kinematic" (PPK).

Depending on the working environment, platform dynamics and receiver quality, two different types of CDGPS solutions can be obtained, i.e., fixed or float solutions [30]. The former is the most accurate one, being able to guarantee up to sub-cm accuracy in the determination of the relative position between the receivers, exploiting the property of carrier-phase ambiguities to become, under suitable measurement combinations and for properly designed receivers, integer numbers. The fixed solution can be robustly generated by processing multi-frequency GPS data and can be obtained, although with reduced time availability, by using single-frequency receivers, which typically rely on the float solution, i.e., they consider carrier-phase ambiguities as real numbers. This is the case of the presented system architecture. Hence, for most of the time epochs, a realistic estimate of the carrier-phase ambiguities can be robustly generated by the adopted single-frequency receivers and the achieved accuracy thus degrades to the order of 10 cm. The error is reduced to a very few cm when fixed solutions are available.

Herein, CDGPS processing is carried out by using the open-source software RTKlib [21]. In particular, the post-processing analysis tool RTKPOST is used, which inputs RINEX observation data and navigation message files (from GPS, GLONASS, Galileo, QZSS, BeiDou and SBAS), and can compute the positioning solutions by various processing modes (such as Single-Point, DGPS/DGNSS, Kinematic, Static, PPP-Kinematic and PPP-Static). In this regard, the "Kinematic" positioning mode is chosen, which corresponds to PPK, with integer ambiguity resolution set to "Fix and hold". RTKPOST outputs the E/N/U coordinates of the flying receiver with respect to the base-station, together with a flag relevant to the solution type (float/fixed). This flag, and the processing residuals, can be used as an estimate of the achieved positioning accuracy.

3. Radar Signal Processing

This section describes the signal processing strategy adopted to process the data collected by the radar system. The various stages of the data processing are summarized in the block diagram of Figure 2. According to this scheme, the input information is the raw radargram (B-scan) collected by the radar, which represents the received radar signal collected at each measurement position (along the flight path) versus the fast-time (i.e., the wave travel time). The final output of the reconstruction procedure is a focused and easily interpretable image depicting the scene under test.

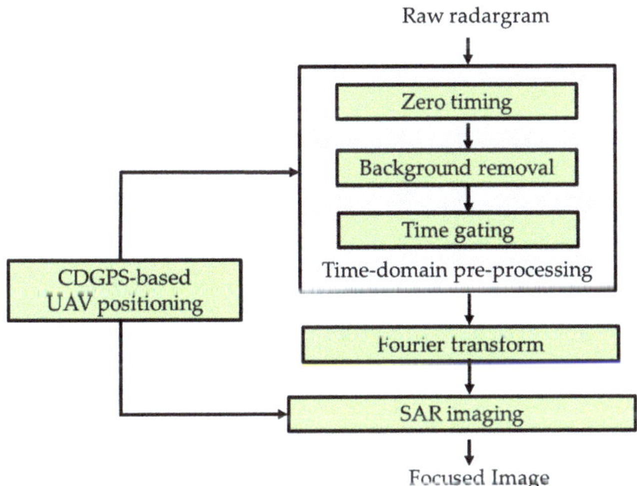

Figure 2. Radar signal processing chain.

As a first stage of the overall reconstruction procedure, a time-domain pre-processing of the radargram is performed by applying the following operations [31–33]:

- Zero-timing;

- Background removal;
- Time-gating.

The zero-timing consists of setting the starting instant of the fast-time axis in such a way that the range of the signal reflected by the air–soil interface at the first measurement point of the flight trajectory is coincident with the UAV flight height estimated by the CDGPS processing.

The background removal is a filtering procedure that allows mitigating the effects of the strong coupling between the transmitting and receiving radar antennas, which is a spatially constant signal. This filter replaces each single radar trace (A-scan) of the radargram with the difference between it and the average of all the traces of the radargram collected along the flight trajectory.

The time-gating procedure selects the interval (along the fast-time) of the radargram, where signals scattered from targets of interest occur. This allows a reduction of environmental clutter and noise effects. Herein, the UAV altitude is exploited to define a suitable time window around the time where reflection of the air–soil interface occurs.

After the time-domain pre-processing stage, each trace in the radargram is transformed into the frequency domain by using the Fast Fourier Transform (FFT) algorithm. Then, the frequency-domain data are processed according to the radar imaging approach detailed in the next subsection.

3.1. Radar Imaging Approach

Let us refer to the 3D scenario sketched in Figure 3. The ultra-wideband radar transceiver onboard the UAV illuminates the scene with transmitting and receiving antennas pointed at nadir (down-looking mode), i.e., at a zero incidence angle with respect to the normal to the air–soil interface. The radar can be considered operating in monostatic mode, since transmitting and receiving antennas have negligible offset in terms of the probing wavelength. At each measurement point along the flight trajectory Γ, the transceiver records the signals scattered from the targets over the angular frequency range $\Omega = [\omega_{min}, \omega_{max}]$. Therefore, multimonostatic and multifrequency data are collected.

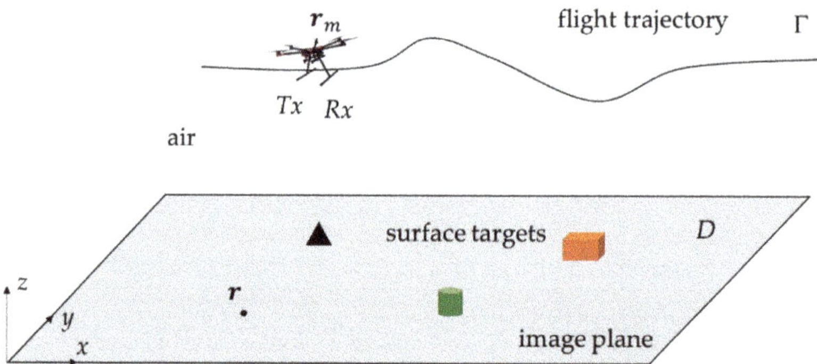

Figure 3. UAV-borne radar imaging scenario.

The trajectory Γ has an arbitrary shape in space and each measurement point is described by the position vector $r_m = x_m\hat{x} + y_m\hat{y} + z_m\hat{z}$. The targets are supposed to be located into the planar investigation domain D, which is coincident with the air–soil interface assumed at $z = 0$. The time dependence $e^{j\omega t}$ is assumed and dropped.

The radar signal model is based on the following assumptions: (i) the antennas have a broad radiation pattern; (ii) the targets are in the far-field region with respect to the radar antennas; (iii) a linear model of the scattering phenomenon is assumed, hence the mutual interactions between the targets are neglected [22]. Accordingly, the scattered signal at each measurement point r_m is expressed by the following linear integral equation [16,34]:

$$E_s(r_m, \omega) = I(\omega) \iint_D \frac{e^{-j2k_0|r_m - r|}}{|r_m - r|^2} \sigma(r) dr = \mathcal{L}\sigma \tag{1}$$

where $I(\omega)$ is the spectrum of the transmitted pulse, $\sigma(r)$ is the unknown reflectivity function at a point $r = x\hat{x} + y\hat{y}$ in D, $k_0 = \omega/c_0$ is the propagation constant in free-space ($c_0 \cong 3 \cdot 10^8$ [m/s] is the speed of light) and $|r_m - r|$ is the distance between the measurement point and the generic point of the investigation domain D. It is worth noting that the spectrum $I(\omega)$ may be assumed unitary within the system bandwidth and therefore, for notation simplicity, it will be omitted.

The linear operator $\mathcal{L}$ maps the space of the unknown object function (reflectivity of the scene) into the space of data (measured scattered field). The reflectivity function $\sigma(r)$ accounts for the difference between the electromagnetic properties of the targets (dielectric permittivity, electrical conductivity) and the free space ones. Accordingly, the targets are searched for as anomalies with respect to the free-space scenario and appear in the "focused image" as the regions where the modulus of the reflectivity function is different from zero.

The radar imaging is faced as the inversion of the linear integral Equation (1) and this is performed by computing the adjoint of the forward scattering operator $\mathcal{L}$ [23]:

$$\tilde{\sigma}(r) = \mathcal{L}^+ E_s = \int_\Gamma \int_\Omega E_s(r_m, \omega) \frac{e^{j2k_0|r_m - r|}}{|r_m - r|^2} dr_m \, d\omega \tag{2}$$

where $\mathcal{L}^+$ is the adjoint operator of $\mathcal{L}$.

The adjoint inversion scheme given by Equation (2) is also referred as frequency-domain back-projection [35], since the measured signal is back projected to the point where it is generated and the image is formed as the coherent summation of these contributions.

The numerical implementation of the inversion is performed by discretizing Equation (2) by applying the Method of Moments [36]. The scattered field is discretized by $M \times N$ data, where M is the number of measurement points (x_m, y_m, z_m), $m = 1, 2 \ldots, M$ and N is the number of angular frequencies ω_n, $n = 1, 2 \ldots N$ sampling the work frequency bandwidth Ω. The domain D is discretized by $P \times Q$ pixels (x_p, y_q), where $p = 1, 2 \ldots P$, and $= 1, 2 \ldots Q$ (see Figure 4). After removing unessential constants, the inversion scheme in Equation (2) is rewritten in discrete form as:

$$\tilde{\sigma}(x_p, y_q) = \sum_{m=1}^{M} \sum_{n=1}^{N} E_s(x_m, y_m, z_m, \omega_m) \frac{e^{\frac{j2\omega_n}{c_0}\sqrt{(x_p - x_m)^2 + (y_q - y_m)^2 + z_m^2}}}{(x_p - x_m)^2 + (y_q - y_m)^2 + z_m^2} \quad \begin{array}{l} p = 1 \ldots P, \\ q = 1 \ldots Q \end{array} \tag{3}$$

According to the assumption of antennas having a broad radiation pattern, Equation (3) sums coherently the multi-frequency data collected along the whole trajectory Γ for each pixel in D. Therefore, the radar image is obtained by computing Equation (3) for all pixels in D and plotting the magnitude of the retrieved reflectivity values normalized with respect to their maximum value.

In this process, the precise measurement positions of the UAV obtained with the CDGPS processing are considered. The exploitation of the positioning information allows accurate images, as already pointed out in the airborne radar imaging context [24,37].

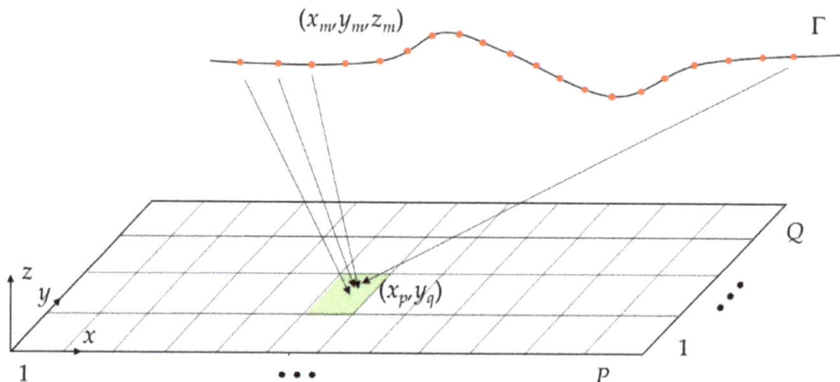

Figure 4. Discretization of the radar imaging problem.

3.2. Resolution Analysis

This subsection aims at investigating the spatial resolution limits of the proposed M-UAV radar imaging system. The analysis covers the effect of the measurement parameters on the resolution limits in the image plane D. To achieve this goal, we compute the point spread function (PSF) of the system, i.e., the reconstruction of a point-like target [23]. For a point-like target located at $r_t = x_t \hat{x} + y_t \hat{y}$ and having unitary reflectivity, the related scattered field is expressed, according to Equation (1), as:

$$E_s(r_m, \omega) = \frac{e^{-j2k_0|r_m - r_t|}}{|r_m - r_t|^2} \tag{4}$$

After plugging Equation (4) in the adjoint inversion formula Equation (2), we get the following expression for the PSF

$$\text{PSF}(r; r_t) = \int_\Gamma \int_\Omega \frac{e^{-j2k_0(|r_m - r_t| - |r_m - r|)}}{|r_m - r_t|^2 |r_m - r|^2} dr_m \, d\omega \tag{5}$$

allowing the evaluation of the resolution as a function of the system parameters and the flight trajectory. Hence, Equation (5) is useful, on one hand, for planning the measurement campaign according to the requirements of the applicative context of interest and, on the other side, for investigating how deviations, with respect to the nominal flight path, affect the achievable imaging performance.

Before proceeding further, it is worth recalling the resolution formulas holding for an ideal rectilinear flight path. These formulas provide useful insight into radar imaging also under non-ideal motion and allow foreseeing, at least in a qualitative way, the effect of the main measurement parameters.

Let us consider the geometry sketched in Figure 5, where the UAV moves at a fixed height h following a rectilinear trajectory directed along the x-axis. The along-track resolution Δ_x is determined by the central frequency f_c of the radar and the maximum view angle θ fixed by half-length of the synthetic aperture [32]:

$$\Delta_x = \frac{c_0}{4 f_c \sin \theta} \tag{6}$$

that in the small angle approximation rewrites as [38]:

$$\Delta_x = \frac{c_0}{4 f_c \theta} \tag{7}$$

The range resolution is related to the radar system bandwidth B by the classical formula [39]:

$$\Delta_r = \frac{c_0}{2B} \tag{8}$$

The across-track resolution Δ_y is evaluated from the projection of the 3D target reconstruction over the image plane (see Figure 5b). If r denotes the target range with respect to the antenna, then the 3D target reconstruction is the cylindrical shell having its axis coincident with the measurement line and its inner and outer radius equal to $r - \Delta_r$ and $r + \Delta_r$, respectively. Note that only a part of the shell is shown in Figure 5b, for sake of clarity. The across-track resolution Δ_y is calculated as the intersection of the cylindrical shell with the image plane $z = 0$ and is given by

$$\Delta_y = \sqrt{d^2 + \Delta_r^2 + 2\Delta_r \sqrt{h^2 + d^2}} - d \tag{9}$$

where d is the across-track distance between the fight trajectory and the target (see Figure 5a,b). According to Equation (9), the across-track resolution gets worse when the UAV flies at a higher altitude h and, when the target is illuminated at nadir ($d = 0$), it turns out that:

$$\Delta_y = \Delta_r \sqrt{1 + \frac{2h}{\Delta_r}} \tag{10}$$

i.e., the across-track resolution is finite and larger than the range resolution Δ_r.

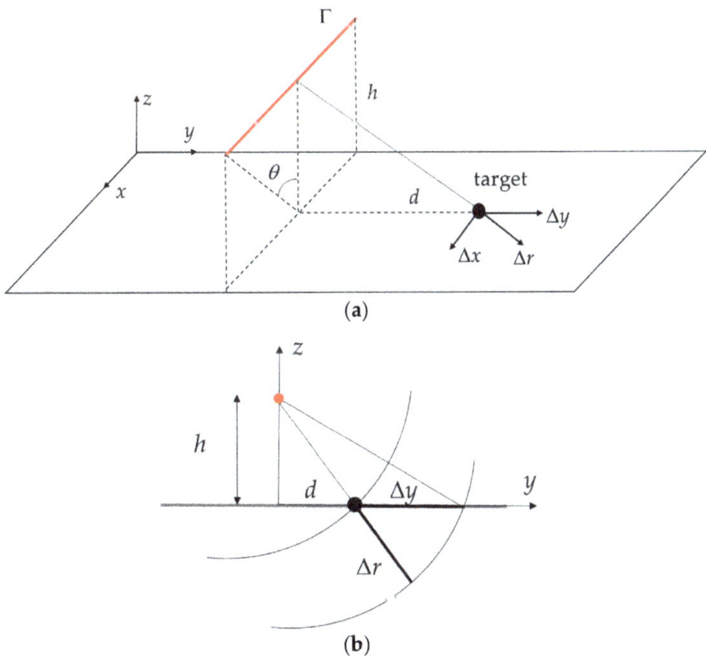

Figure 5. Radar imaging with an ideal rectilinear flight path: (**a**) 3D view; (**b**) view in the y–z plane.

Equation (9) also reveals that, for a fixed value of h and Δ_r, Δ_y improves as long as the target moves away from the measurement line. Most notably, the asymptotic value of the across-track resolution is found as d approaches to infinity.

$$\Delta_y = \lim_{d \to \infty} \sqrt{d^2 + \Delta_r^2 + 2\Delta_r \sqrt{h^2 + d^2}} - d = \Delta_r \tag{11}$$

Based on the results in the Equations (10) and (11), the following inequality holds:

$$\Delta_r \le \Delta_y \le \Delta_r \sqrt{1 + \frac{2h}{\Delta_r}} \tag{12}$$

Note that if the system bandwidth B goes to zero, the range resolution Δ_r becomes infinite and it is no longer possible to resolve targets along the direction perpendicular to the track, as already noticed in [24].

Figure 6 depicts the across-track resolution Δ_y as a function of the target offset d and the flight altitude h. The contour plot has been produced by applying Equation (9) and considering the bandwidth of the radar system (i.e., $B = 1.7$ GHz) introduced in Section 2.

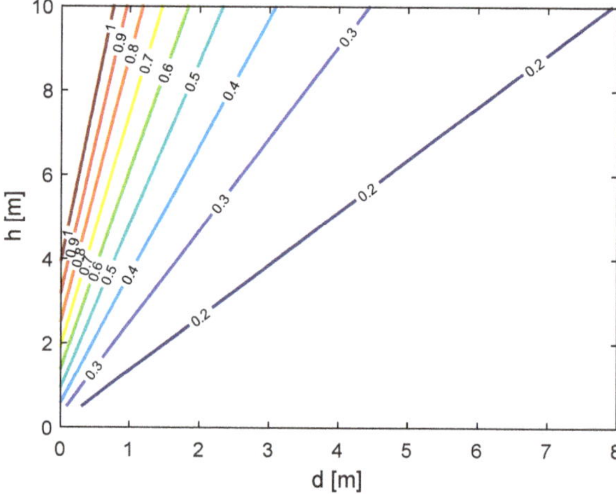

Figure 6. Contour plot of the across-track resolution Δ_y versus d and h, expressed in meters, in the case of a rectilinear flight trajectory.

As previously pointed out, the resolution degrades when increasing the flight altitude h for a fixed value of d or when reducing d for a fixed value of h.

Figure 7 provides an example of the PSF computed according to Equation (5) by considering an investigation domain $D = [-3, 3] \times [-3, 3]$ m^2, which is discretized by square image pixels with size 0.01 m, and two different values of the target offset d (i.e., $d = 0$ m and $d = 2$ m). The scattered field data are sampled evenly with 0.01 m step along the trajectory Γ at a flight altitude $h = 5$ m.

Figure 7a,b reports the PSF reconstruction for the case of a rectilinear trajectory covering the interval $[-3, 3]$ along x.

Figure 7c,f considers the effect of a non-rectilinear UAV flight trajectory; specifically, the x-directed rectilinear trajectory of Figure 7a,b is perturbed in the $x - y$ plane and modified in accordance with the co-sinusoidal function

$$y = \pm 0.15 \; \cos\left(\frac{\pi x}{12}\right) \tag{13}$$

Equation (13) defines a curved trajectory over the interval $[-3, 3]$ m with a maximum deviation of 0.15 m along y with respect to the rectilinear trajectory.

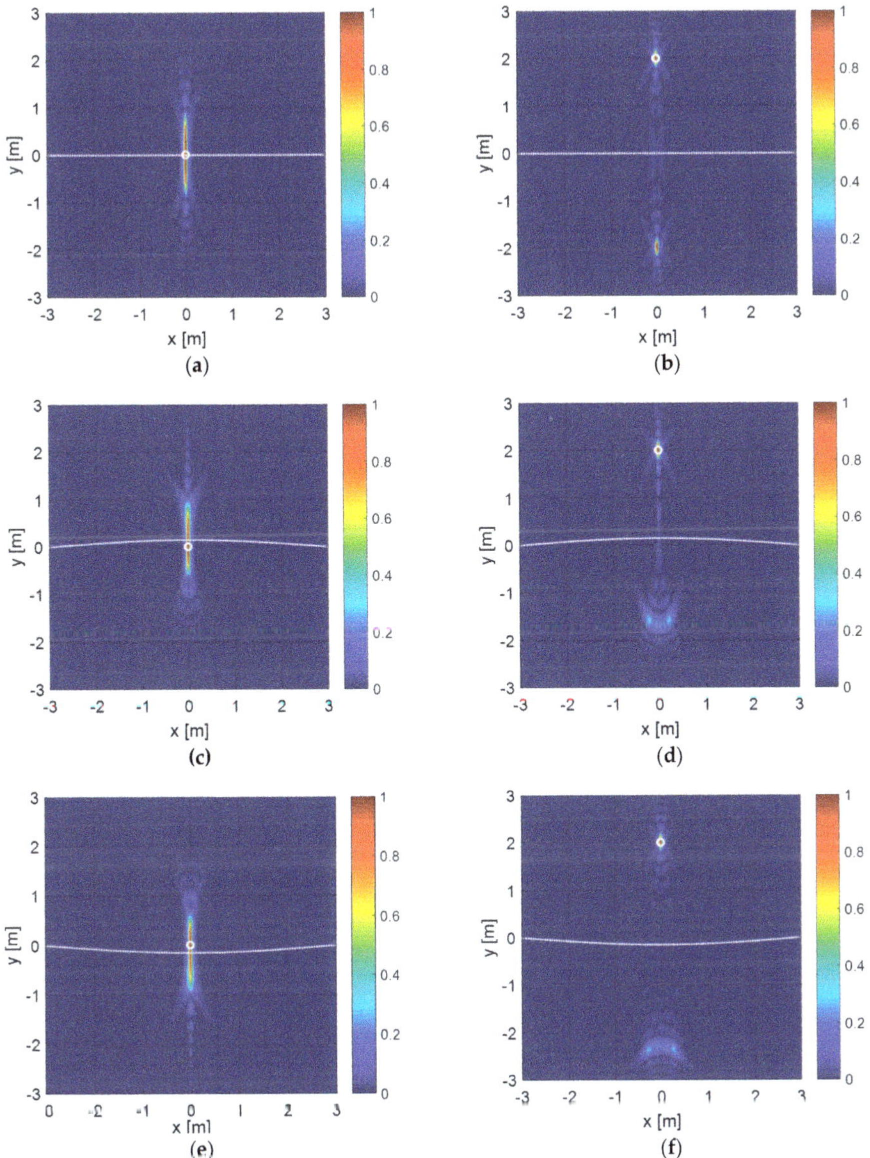

Figure 7. Point spread function (PSF) amplitude for a flight altitude $h = 5\ m$: ideal rectilinear trajectory and point like target with offset: (**a**) $d = 0$ m, (**b**) $d = 2$ m. Curved trajectory as described by Equation (13) and point-like target located with offset: (**c,e**) $d = 0$ m, (**d,f**) $d = 2$ m. The white dashed line represents the trajectory and the white circle denotes the target.

Figure 7a,b shows that a focused spot along and across the track is obtained in correspondence of the target position and the along-track resolution does not change when the target is located at the radar nadir ($d = 0$ m) or at the point $(0, 2)$ m. Conversely, the across-track resolution improves when the target is far from the nadir, as predicted by Equation (9). However, in this latter case, a false target appears at the specular position with respect to the flight trajectory, i.e., at $(0, -2)$ m. This phenomenon is the so-called left–right ambiguity [40] and is due to the radar's inability to discriminate left $(y > 0)$ and right $(y < 0)$ targets located at the same distance with respect to the measurement line.

In addition, Figure 7c,f shows that, as expected, even with a slight trajectory deviation with respect to the rectilinear path, the PSF is no longer symmetric with respect to the trajectory. Most notably, when the target is placed at $(0, 2)$ m (see Figure 7d,f), the false target due to the left–right ambiguity appears distorted and with a lower intensity, with respect to Figure 7b. Indeed, when the trajectory is not rigorously rectilinear, the left and right targets are in some way discriminated by the radar because their echoes have different propagation delays at each measurement point. However, the beneficial effect provided by the trajectory curvature in mitigating the false target becomes less relevant when the flight altitude h increases, since left and right targets produce scattering signals with "more similar" propagation delays. This statement is corroborated by the images in Figure 8a,b, which are analogous to Figure 7c,d but for the altitude that is $h = 10$ m. As expected, by increasing flight altitude, the resolution across-track degrades regardless of the position of the target and the left–right ambiguity problem turns out to be more evident. The amplitude of the false target seen in Figure 8b is, indeed, stronger compared to the one observed in Figure 7d,f.

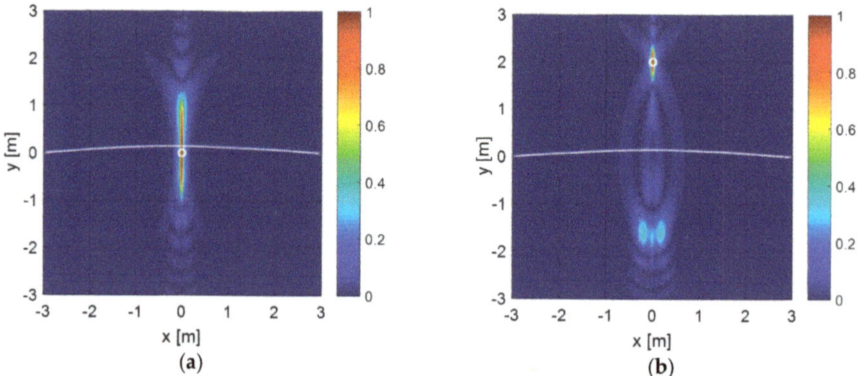

Figure 8. PSF amplitude for $h = 10$ m and a curved trajectory: (**a**) point like target at $d = 0$; (**b**) point like target at $d = 2$. The white dashed line shows the trajectory; the white circle denotes the target.

The along- and across-track resolution values referred to the considered numerical examples are summarized in Table 1.

Table 1. Along- and across-track resolution values.

	Along-Track Resolution (m)	Across-Track Resolution (m)
Rectilinear path, target offset $d = 0$ m, flight altitude $h = 5$ m	0.04	0.95
Rectilinear path, target offset $d = 2$ m, flight altitude $h = 5$ m	0.04	0.25
Path $y = 0.15\cos\left(\frac{\pi x}{12}\right)$, target offset $d = 0$ m, flight altitude $h = 5$ m	0.04	0.95
Path $y = 0.15\cos\left(\frac{\pi x}{12}\right)$, target offset $d = 2$ m, flight altitude $h = 5$ m	0.04	0.25
Path $y = -0.15\cos\left(\frac{\pi x}{12}\right)$, target offset $d = 0$ m, flight altitude $h = 5$ m	0.04	0.95
Path $y = -0.15\cos\left(\frac{\pi x}{12}\right)$, target offset $d = 2$ m, flight altitude $h = 5$ m	0.04	0.25
Path $y = 0.15\cos\left(\frac{\pi x}{12}\right)$, target offset $d = 0$ m, flight altitude $h = 10$ m	0.07	1.30
Path $y = 0.15\cos\left(\frac{\pi x}{12}\right)$, target offset $d = 2$ m, flight altitude $h = 10$ m	0.07	0.47

4. Experimental Results

The M-UAV radar imaging system has been experimentally tested at an authorized site for amateur UAV testing flights in Acerra, Naples, Italy. The experiment aimed at testing the ability of the

CDGPS technique to estimate the UAV position with the accuracy required for target imaging and, thus, to verify the capability of the overall radar imaging system. The experiment was carried out during a sunny day with a weak wind state. Two metallic trihedral corner reflectors, having a size $D = 0.40\ \text{m} \times 0.40\ \text{m} \times 0.57\ \text{m}$ and referred as Target 1 and Target 2, were used as on-ground targets placed at a relative distance of 10 m one from the other along the flight direction; one of them (i.e., Target 2) was covered with a cardboard box (see Figure 9).

Figure 9. The UAV-borne radar imaging scenario.

The UAV was manually piloted and two surveys at different altitudes, in the following referred to as Track 1 and Track 2, were carried out. Both tracks were performed on the same scenario by positioning the UAV nearly at the same starting point (x, y). Track 1 had a duration of 17.5 s and covered a path 31.4 m long at an average altitude $h = 4$ m; along this track, data were gathered at 251 unevenly spaced measurement points. Track 2 had a duration of 21.7 s and covered a 33 m long path at an average altitude $h = 10$ m; along this track, data were gathered at 331 unevenly spaced measurement points. The radar parameters set for the data acquisition are summarized in Table 2. Note that we considered flight altitude values in the range 5–10 m to operate with a suitable signal-to-noise ratio. Indeed, a major constraint in our system is the limited transmit power of the radar, whose maximum level is declared to be approximately –13 dBm by the manufacturer.

Table 2. Radar system parameters.

Parameters	Specification
Carrier Frequency	3.95 GHz
Frequency Band	1.7 GHz
Maximum Emitted Power	−13 dBm
Maximum Dynamic Range	75 dB
Pulse Repetition Frequency	14.28 Hz
Received Signal Sampling Frequency	16 GHz

The raw radargrams, i.e., the data collected during the two surveys, are depicted in Figure 10a,b while the filtered radargrams (after the time domain pre-processing stage) are given in Figure 11a,b. It is worth pointing out that the horizontal axis shows the slow-time, i.e., the duration of the flight in seconds, while the vertical axis is the fast-time, i.e., the observation time window during which the data are gathered for each radar position, once that the time-zero correction has been performed. The fast-time is expressed in nanoseconds. The white dotted line represents the air/soil interface achieved by converting the variable UAV flight altitude h estimated by the CDGPS into an equivalent travel time t_h by using the formula $t_h = 2h/c_0$.

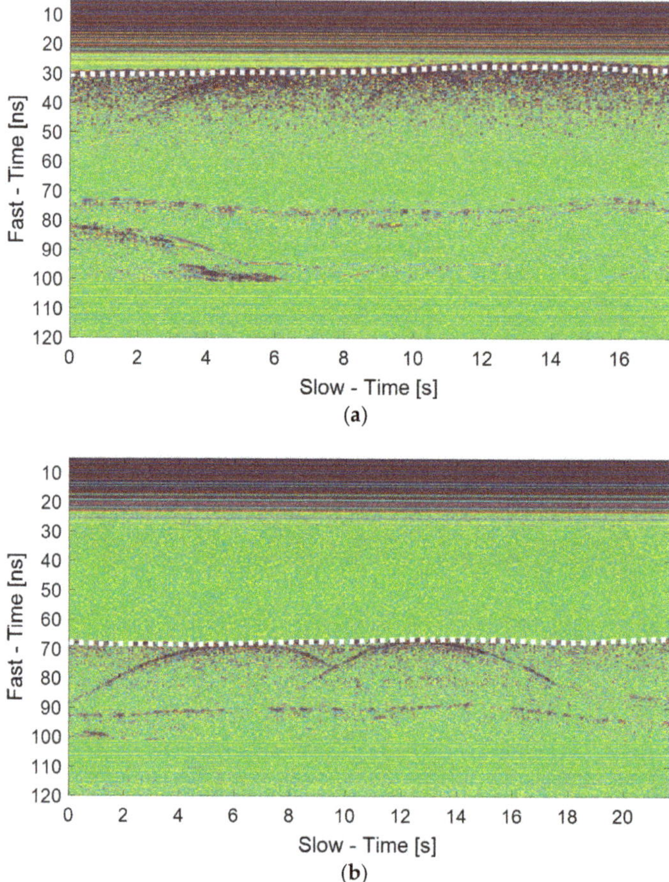

Figure 10. Raw radargrams: (**a**) Track 1; (**b**) Track 2. The white dotted line represents the variable UAV flight altitude h estimated by the Carrier-Phase Differential GPS (CDGPS) and transformed into the equivalent travel time by: $t_h = 2h/c_0$.

From Figures 10 and 11, one can observe that the CDGPS provides an accurate estimation of the flight altitudes and the targets' responses are visible as hyperbolas whose apex occurs at the fast-time where nadir surface reflection is observed. Moreover, Figure 10a,b shows that clutter signals, due to metallic awnings located on the entry side of the flight site, appear at fast-times greater than 70 ns in Figure 10a and 90 ns in Figure 10b. These undesired signals, as well as the mutual coupling between transmitting and receiving antennas, are removed by a time-domain pre-processing (see Figure 11a,b). The filtered radargrams have been obtained by performing the background removal and setting as fast-time gating window the portion occurring 6 ns before and 24 ns after the air–soil interface response seen at nadir. The filtered data have been transformed into the frequency domain by sampling the radar bandwidth [3.1, 4.8] GHz into 341 evenly spaced frequency samples and have been processed according to the inversion procedure described in Section 3.1.

Before showing the focused radar images, we provide quantitative data about the positioning accuracy of the UAV. Specifically, Table 3 summarizes the maximum positioning errors achieved with the CDGPS technique along Tracks 1 and 2. These errors are the standard deviations provided by the RTKlib tool, which measure the positioning errors along the three coordinate axes based on a priori error models and error parameters [21]. The maximum errors in the horizontal plane are always smaller than the error along z, which is 9.4 cm in the worst case (Track 2).

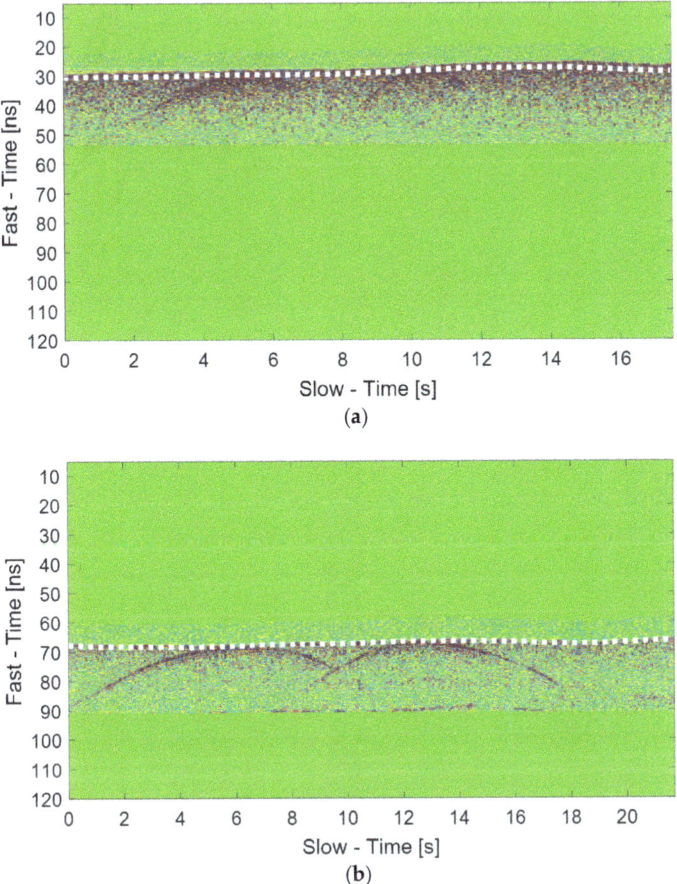

Figure 11. Filtered radargrams: (**a**) Track 1; (**b**) Track 2. The white dotted line represents the variable UAV flight altitude h estimated by the CDGPS and transformed in the equivalent travel time by: $t_h = 2h/c_0$.

Table 3. Maximum errors of the CDGPS technique.

Errors (cm)	x	y	z
Track 1	4.3	4.6	9.4
Track 2	0.6	0.8	1.5

The focused images of the surveyed scenario are depicted in Figure 12a,b for Track 1 and Track 2, respectively. These images have been obtained by considering a square planar investigation domain D at $z = 0$ m, whose origin corresponds to the starting point of the UAV tracks into the x–y plane and whose side is 18 m. The domain D has been evenly discretized by pixels having side 0.01 m.

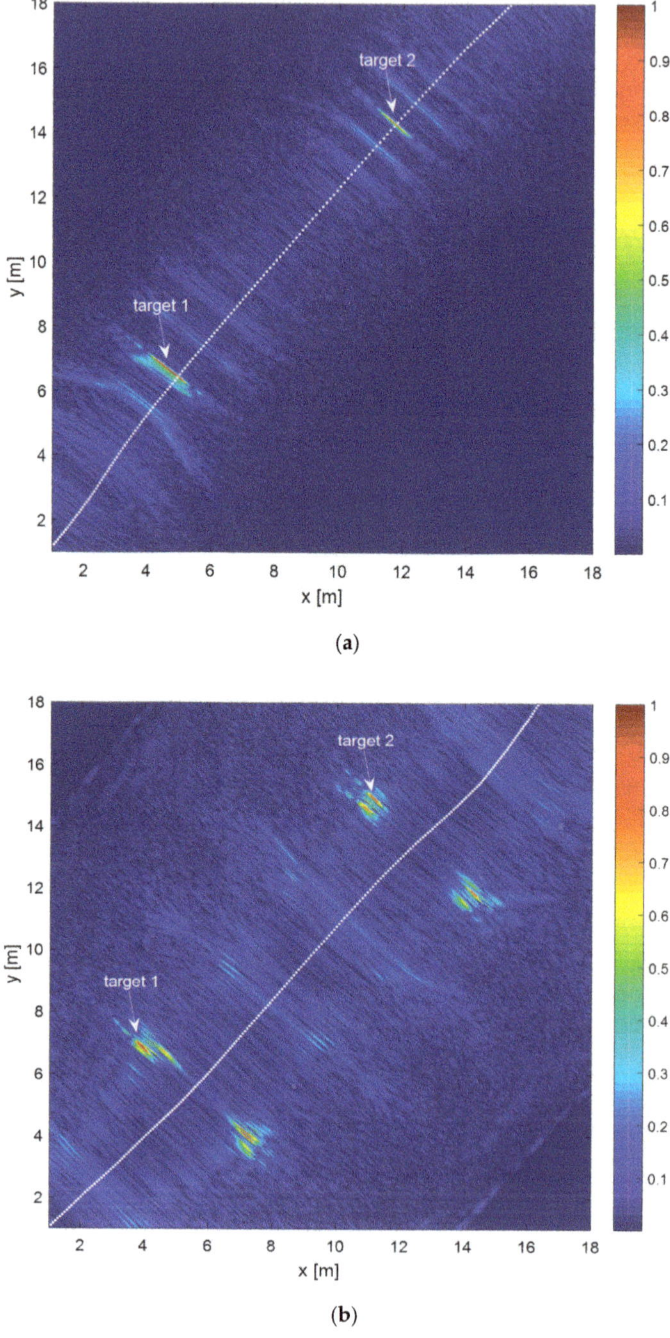

Figure 12. Focused image of the scenario under test: (**a**) Track 1; (**b**) Track 2. The dotted white line represents the flight path as projected onto the investigated domain.

In Figure 12a,b, the dotted white line represents the M-UAV trajectory as estimated by the CDGPS and projected onto the investigated domain. According to the analysis presented in Section 3.2, Figure 12a shows that when the targets are illuminated at nadir, i.e., when the distance d approaches to zero, single spots appear and no ambiguities occur. Conversely, false targets due to the left–right ambiguity problem appear when the UAV flight path does not cover the targets (see Figure 12b). However, coherently with the PSFs shown in Figure 8b, the false targets appear slightly distorted and with lower intensity compared to the real target reconstructions owing to the trajectory curvature. As a result, it is possible to discriminate the actual targets from the ambiguous ones. Table 4 reports the experimental along- and across-track resolution values as estimated by Figure 12a,b for both targets. For comparison, the table reports the theoretical resolution values referred to a rectilinear flight path at the average altitudes $h = 4$ m and $h = 10$ m. The experimental and theoretical resolution values are quite consistent. Notably, the experimental along-track resolution decreases slightly when the flight altitude increases and the target offset d is not null, while the across-track one improves when d increases. It is worth pointing out that the corner reflectors emphasize the radar echoes but they are not actually ideal point targets. Consequently, some discrepancies on resolution values are expected and this outcome is confirmed by the comparison between the experimental and theoretical data reported in Table 4.

Table 4. Along- and across-track resolution values.

		Experimental Along-Track Resolution (m)	Theoretical Along-Track Resolution (m)	Experimental Across-Track Resolution (m)	Theoretical Across-Track Resolution (m)
Track 1[1]	Target 1	0.05	0.02	1.15	0.85
	Target 2	0.04	0.02	0.99	0.85
Track 2[2]	Target 1	0.07	0.03	0.51	0.41
	Target 2	0.09	0.03	0.50	0.41

[1] $h = 4m$, $d = 0m$; [2] $h = 10m$, $d = 2m$

5. Discussion

This work deals with a feasibility study on small UAV-based radar imaging when the scene under investigation is probed with a single measurement line and the imaging domain is a plane at a fixed altitude. The considered acquisition geometry is the simplest one and its achievable imaging capabilities have been studied in Section 4. Regarding the along-track resolution, this parameter is influenced by the maximum illumination angle, which in turn depends on the flight altitude and the length of the synthetic aperture. The flight height and the horizontal displacement between the target and the UAV, instead, influence the across-track resolution. Targets far from the radar nadir are generally better resolved across-track than those seen at nadir; however, an inherent limitation in the imaging arises due to the left–right ambiguity problem. This phenomenon is partially mitigated in the presence of horizontal deviations of the UAV with respect to the ideal rectilinear trajectory. Additionally, flying at a higher altitude can be convenient to enlarge the area of coverage but such choice generally produces a worsening of the spatial resolution both along- and across-track.

A further point worth to be discussed concerns the inability of the present imaging configuration to provide unambiguous and high-resolution 3D target reconstructions. To clarify this point, it is useful to refer to Figure 13 showing how the reconstruction of the target changes when the image plane is not the correct one. In particular, in Figure 13, we show how a point target located on the plane D_0 at $z = 0$ is imaged on three planes D_0, D_1, D_2 placed at different altitudes, i.e., $z = 0$, $z = z_1$, and $z = z_2$.

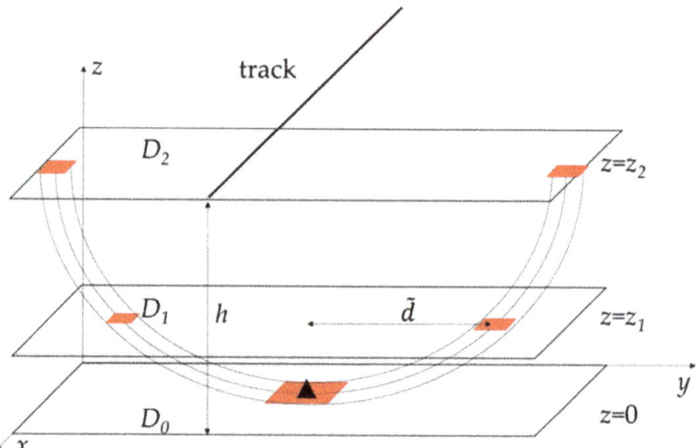

Figure 13. Reconstruction of a point target on image planes at different elevations. The true target (black triangle) is illuminated at the radar nadir and its reconstruction is represented by the red rectangles.

If the image plane coincides with the plane where the target is located, i.e., D_0, the target is reconstructed at the correct position. When the image plane is different from D_0, i.e., D_1 or D_2, due to the cylindrical symmetry of the 3D target reconstruction, the target is imaged at a position different from the true one in the considered plane. The position of the reconstructed target is equal to the intersection point between the 3D reconstruction and the plane where the imaging is carried out. Furthermore, due to the left–right ambiguity, two specular targets appear on both sides of the track (see red rectangles on planes D_1 or D_2). The spatial offset $\tilde{d}$ in the x–y plane between the true target and the reconstructed one for an image plane at a height z can be derived after straightforward geometrical considerations and is given by

$$\tilde{d} = \sqrt{d^2 + 2hz - z^2} - d. \tag{14}$$

This last formula holds also in the more general case when the target is not illuminated at the radar nadir, as in Figure 13, and d is the horizontal distance between the target and the track.

The geometry in Figure 13 also reveals that the target can be detected (but not correctly localized) when the imaging plane is placed at a higher elevation with respect to the target. Indeed, in this case, it is still possible to find two intersection points between the 3D target reconstruction and the image plane. Conversely, the target cannot be identified at all when it is located above the image plane since this last no longer intersects the 3D target reconstruction.

A numerical example showing the effect of the elevation of the image plane is presented in the case of a multi-target scenario. Specifically, the example refers to the rectilinear trajectory and simulation parameters already considered in Section 3.2. The scene comprises three point targets T1, T2, T3 aligned along the flight track and located at coordinates: $(-2, 0, 0)$ m, $(0, 0, 0.2)$ m, $(2, 0, 0.4)$ m. The reconstructions results achieved on three images planes at $z = 0, 0.2$ and 0.4 m are displayed in Figure 14a–c, respectively. As can be observed in Figure 14a, only the target T1 is imaged and correctly localized in the plane $z = 0$ m while the targets T2 and T3 are not detected because they are located above the image plane. When the image plane is fixed at $z = 0.2$ m, the target T2 is the only one to be correctly localized while T1 is imaged a different location with a spatial offset with respect to the true position. The target T3 is still not detectable because its elevation is greater than the height of the image plane. Finally, Figure 14c shows the reconstruction in the plane $z = 0.4$m. In this case, all targets are detected but only T3 is correctly localized.

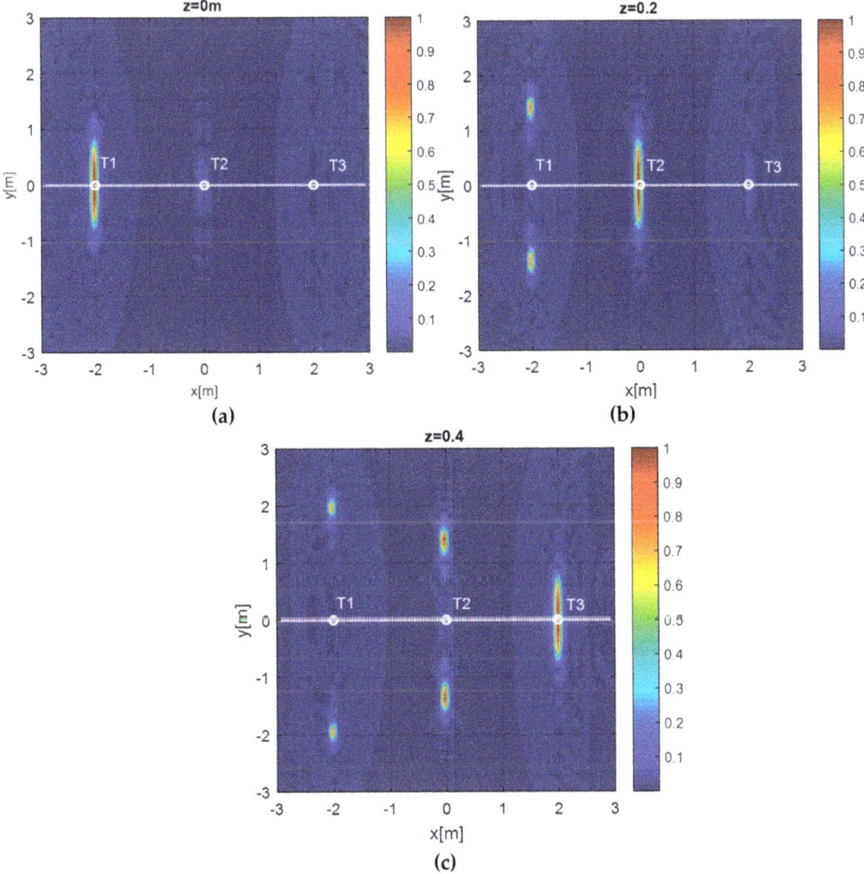

Figure 14. Reconstruction results in a three-target scenario (**a**) image plane at $z = 0$ m; (**b**) image plane at $z = 0.2$ m; (**c**) image plane at $z = 0.4$ m.

Table 5 reported below compares the true and reconstructed targets' positions achieved in each image plane. The maximum of each spot in the images of Figure 14a–c is considered as the estimate of the targets' positions. Note that the ± sign appears in the presence of the left–right ambiguity problem.

Table 5. Estimated and true target positions for different imaging planes.

	True Target Position	Retrieved Target Positions Versus Height of Image Plane		
		$z = 0$ m	$z = 0.2$ m	$z = 0.4$ m
T1	$(-2, 0, 0)$ m	$(-2, 0, 0)$ m	$(-2, \pm 1.4, 0.2)$	$(-2, \pm 1.99, 0.4)$ m
T2	$(0, 0, 0.2)$ m	-	$(0, 0, 0.2)$	$(0, \pm 1.4, 0.4)$ m
T3	$(2, 0, 0.4)$ m	-	-	$(2, 0, 0.4)$ m

An improvement of the approach in terms of resolution and left–right ambiguity suppression toward a high-resolution 3D imaging can be achieved by collecting wideband scattered field data along multiple (parallel) measurement tracks. A similar measurement configuration has been recently studied in the single-frequency case [24]. The theoretical and experimental assessment of such a configuration in the multifrequency case will be the subject of future research.

As a further upgrade of the radar imaging system, the possibility of using a gimbal, as suggested in [41,42], will be considered to achieve major flexibility in the data acquisition.

6. Conclusions

A proof of concept of a Multicopter Unmanned Aerial Vehicle (M-UAV) radar imaging system has been developed by integrating a miniaturized and commercial radar system onboard a small M-UAV. The imaging system has been equipped with two Global Positioning System (GPS) receivers, the first one located onboard the M-UAV platform, and, the second one used as a ground-based station with the aim of exploiting the Carrier-Phase-Based Differential GPS (CDGPS) technique. This latter allows estimating the 3D M-UAV flight path with centimeter accuracy. Moreover, an advanced imaging approach, based on the adjoint inverse scattering problem, has been adopted to obtain focused images of on surface targets in the case of a single flight track. This approach exploits the 3D M-UAV trajectory estimate provided by the CDGPS into the reconstruction stage.

A theoretical/numerical analysis has been preliminary conducted to evaluate the effect of the overall system and measurement configuration parameters on the imaging system performance. In addition, a proof of concept measurement campaign has been performed. The flight tests have been carried out by manually piloting the UAV at an authorized site for amateur and the experimental results have demonstrated the capability of the system to obtain very good imaging results, comparable to those foreseen with the theoretical analysis. This was possible thanks to the accurate UAV positioning estimation, which means an accurate knowledge of the measurement points, that is a key factor for a reliable focusing of the targets. It is worth noting that, despite the simple and light radar system adopted in this work, the necessity of dealing with high frequency a working band and centimetric probing wavelength (7.5 cm at the frequency of 4 GHz) has made significantly challenging the necessity to have an accurate UAV positioning estimate. This was necessary for a reliable focusing procedure requiring knowledge of the location of the measurement points along the flight trajectory with an accuracy comparable with the probing wavelength.

A final comment is dedicated to future developments. Indeed, to overcome the ambiguity effects caused by the nadir antenna pointing, non-rectilinear trajectory, such as circular or slanting flights are worth being considered. The planning of a measurement campaign involving this kind of flight is the subject of current work. In addition, further flight tests will be conducted to assess the subsurface imaging system capability. Moreover, waypoint following and grid surveys will be exploited to regularly sample the area of interest, and multi-constellation/multi-frequency GNSS will be tested. In this frame, more sophisticated flight/navigation modes and 3D tomographic imaging approaches based on multiple measurement lines will be exploited, in order to open novel remote sensing perspectives in structural monitoring and cultural heritage contexts.

Author Contributions: Conceptualization, I.C., G.G. and F.S.; Methodology, I.C., G.G., F.S. and G.F.; Software and Validation, I.C., G.G., C.N., G.E. and G.F.; Formal Analysis, I.C., G.G., F.S.; Investigation, I.C., C.N., G.E. and G.F.; Data Curation, I.C., C.N., G.E. and G.F.; Writing-Original Draft Preparation, I.C., G.G., G.L., C.N., G.E., A.R. and G.F.; Supervision, F.S.; Project Administration, I.C. and F.S.; Funding Acquisition, I.C. and F.S. All authors have read and agreed to the published version of the manuscript.

Funding: This research received no external funding.

Acknowledgments: This research was funded by CAMPANIA FESR Operational Program 2014-2020, under the VESTA project "Valorizzazione E Salvaguardia del paTrimonio culturAle attraverso l'utilizzo di tecnologie innovative" (Enhancement and Preservation of the Cultural Heritage through the use of innovative technologies VESTA), grant number CUP B83D18000320007.

Conflicts of Interest: The authors declare no conflict of interest.

References

1. Everaerts, J. The Use of Unmanned Aerial Vehicles (UAVs) for Remote Sensing and Mapping. *Int. Arch. Photogramm. Remote Sens. Spat. Inf. Sci.* **2008**, *37*, 1187–1192.

2. Quan, Q. *Introduction to Multicopter Design and Control*; Springer: Singapore, 2017.

3. Colomina, I.; Molina, P. Unmanned Aerial Systems for Photogrammetry and Remote Sensing: A review. *ISPRS J. Photogramm. Remote Sens.* **2014**, *92*, 79–97. [CrossRef]

4. Yao, H.; Qin, R.; Chen, X. Unmanned Aerial Vehicle for Remote Sensing Applications—A review. *Remote Sens.* **2019**, *11*, 1443. [CrossRef]

5. Catapano, I.; Di Napoli, R.; Soldovieri, F.; Bavusi, M.; Loperte, A.; Dumoulin, J. Structural Monitoring via Microwave Tomography-Enhanced GPR: The Montagnole test site. *J. Geophys. Eng.* **2012**, *9*, 100–107. [CrossRef]

6. Oriot, H.; Cantalloube, H. Circular SAR imagery for urban remote sensing. In Proceedings of the 7th European Conference on Synthetic Aperture Radar, Friedrichshafen, Germany, 2–5 June 2008.

7. Chao, H.; Cao, Y.; Chen, Y. Autopilots for Small Unmanned Aerial Vehicles: A survey. *Int. J. Control Autom. Syst.* **2010**, *8*, 36–44. [CrossRef]

8. Massonnet, D.; Souyris, J.C. *Imaging with Synthetic Aperture Radar*; EPFL Press: Lausanne, Switzerland, 2008.

9. Cunliffe, A.M.; Brazier, R.E.; Anderson, K. Ultra-fine grain landscape-scale quantification of dryland vegetation structure with drone-acquired structure-from-motion photogrammetry. *Remote Sens. Environ.* **2016**, *183*, 129–143. [CrossRef]

10. Li, C.J.; Ling, H. High-resolution, downward-looking radar imaging using a small consumer drone. In Proceedings of the 2016 IEEE International Symposium on Antennas and Propagation (APSURSI), Fajardo, Puerto Rico, 26 June–1 July 2016; pp. 2037–2038.

11. Bhardwaj, A.; Sam, L.; Akansha, L.; Martin-Torres, F.J.; Kumar, R. UAVs as remote sensing platform in glaciology: Present applications and future prospects. *Remote Sens. Environ.* **2016**, *175*, 196–204. [CrossRef]

12. Remy, M.A.; de Macedo, K.A.; Moreira, J.R. The first UAV-based P-and X-band interferometric SAR system. In Proceedings of the 2012 IEEE International Geoscience and Remote Sensing Symposium (IGARSS), Munich, Germany, 22–27 July 2012; pp. 5041–5044.

13. Llort, M.; Aguasca, A.; Lopez-Martinez, C.; Martínez-Marin, T. Initial Evaluation of SAR Capabilities in UAV Multicopter Platforms. *IEEE J. Sel. Top. Appl. Earth Obs. Remote Sens.* **2018**, *11*, 127–140. [CrossRef]

14. Amiri, A.; Tong, K.; Chetty, K. Feasibility study of multi-frequency ground penetrating radar for rotary UAV platforms. In Proceedings of the IET International Conference on Radar Systems, Glasgow, UK, 22–25 October 2012; pp. 1–6.

15. Fernández, M.G.; López, Y.Á.; Arboleya, A.A.; Valdés, B.G.; Vaqueiro, Y.R.; Andrés, F.L.H.; García, A.P. Synthetic aperture radar imaging system for landmine detection using a ground penetrating radar on board a unmanned aerial vehicle. *IEEE Access* **2018**, *6*, 45100–45112. [CrossRef]

16. Soumekh, M. *Synthetic Aperture Radar Signal Processing*; Wiley: New York, NY, USA, 1999; Volume 7.

17. Ludeno, G.; Catapano, I.; Renga, A.; Vetrella, A.; Fasano, G.; Soldovieri, F. Assessment of a micro-UAV system for microwave tomography radar imaging. *Remote Sens. Environ.* **2018**, *212*, 90–102. [CrossRef]

18. Chao, H.; Gu, Y.; Napolitano, M. A Survey of Optical Flow Techniques for Robotics Navigation Applications. *J. Intell. Robot. Syst.* **2013**, *73*, 361–372. [CrossRef]

19. Fletcher, I.; Watts, C.; Miller, E.; Rabinkin, D. Minimum entropy autofocus for 3D SAR images from a UAV platform. In Proceedings of the 2016 IEEE Radar Conference (RadarConf), Philadelphia, PA, USA, 2–6 May 2016; pp. 1–5.

20. Kaplan, E.; Hegarty, C.J. *Understanding GPS–Principles and Applications*, 2nd ed.; Artech House: Boston, MA, USA, London, UK, 2006.

21. Available online: http://www.rtklib.com/rtklib_document.htm (accessed on 28 February 2020).

22. Chew, W.C. *Waves and Fields in Inhomogeneous Media*; IEEE Press: Piscataway, NJ, USA, 1995.

23. Bertero, M.; Boccacci, P. *Introduction to Inverse Problems in Imaging*; CRC Press: Boca Raton, FL, USA, 1998.

24. Gennarelli, G.; Catapano, I.; Soldovieri, F. Reconstruction capabilities of down-looking airborne GPRs: The single frequency case. *IEEE Trans. Comput. Imaging* **2017**, *3*, 917–927. [CrossRef]

25. Available online: https://www.humatics.com/products/scholar-radar/ (accessed on 28 February 2020).

26. Standard, G.S.P. Available online: https://trade.ec.europa.eu/tradehelp/standard-grp (accessed on 28 February 2020).

27. Milbert, D. Dilution of precision revisited. *Navigation* **2008**, *55*, 67–81. [CrossRef]

28. Farrell, J.A. *Aided Navigation: GPS with High Rate Sensors*; Mc Graw. Hill: New York, NY, USA, 2–5 June 2008.

29. Groves, P.D. *Principles of GNSS, Inertial, and Multisensor Integrated Navigation Systems*; Artech House: Norwood, MA, USA, 2008.

30. Renga, A.; Fasano, G.; Accardo, D.; Grassi, M.; Tancredi, U.; Rufino, G.; Simonetti, A. Navigation facility for high accuracy offline trajectory and attitude estimation in airborne applications. *Int. J. Navig. Obs.* **2013**, 1–13. [CrossRef]

31. Daniels, D.J. Ground penetrating radar. In *Encyclopedia of RF and Microwave Engineering*; John Wiley & Sons: Hoboken, NJ, USA, 2005.

32. Persico, R. *Introduction to Ground Penetrating Radar: Inverse Scattering and Data Processing*; John Wiley & Sons: Hoboken, NJ, USA, 2014.

33. Catapano, I.; Gennarelli, G.; Ludeno, G.; Soldovieri, F.; Persico, R. Ground-Penetrating Radar: Operation Principle and Data Processing. *Wiley Encycl. Electr. Electron. Eng.* **2019**, 1–23.

34. Balanis, C.A. *Advanced Engineering Electromagnetics*; John Wiley & Sons: Hoboken, NJ, USA, 1999.

35. Solimene, R.; Catapano, I.; Gennarelli, G.; Cuccaro, A.; Dell'Aversano, A.; Soldovieri, F. SAR Imaging Algorithms and some Unconventional Applications: A Unified Mathematical Overview. *IEEE Signal Proc. Mag.* **2014**, *31*, 90–98. [CrossRef]

36. Harrington, R.F. *Field Computation by Moment Methods*; Wiley-IEEE Press: Hoboken, NJ, USA, 1993.

37. Gennarelli, G.; Catapano, I.; Ludeno, G.; Noviello, C.; Papa, C.; Pica, G.; Soldovieri, F.; Alberti, G. A low frequency airborne GPR System for Wide Area Geophysical Surveys: The Case Study of Morocco Desert. *Remote Sens. Environ.* **2019**, *233*, 111409. [CrossRef]

38. Scherzer, O. *Handbook of Mathematical Methods in Imaging*; Springer Science & Business Media: Berlin, Germany, 2010.

39. Richards, M.A. *Fundamentals of Radar Signal Processing*; Tata McGraw-Hill Education: New York, NY, USA, 2005.

40. Cheney, M.; Borden, B. *Fundamentals of Radar Imaging*; Siam: Philadelphia, PA, USA, 2009; Volume 79.

41. Gašparović, M.; Jurjević, L. Gimbal influence on the stability of exterior orientation parameters of UAV acquired images. *Sensors* **2017**, *17*, 401. [CrossRef] [PubMed]

42. Pérez, J.A.; Gonçalves, G.R.; Rangel, J.M.G.; Ortega, P.F. Accuracy and effectiveness of orthophotos obtained from low cost UASs video imagery for traffic accident scenes documentation. *Adv. Eng. Softw.* **2019**, *132*, 47–54. [CrossRef]

 *remote sensing* 

Article

The ASI Integrated Sounder-SAR System Operating in the UHF-VHF Bands: First Results of the 2018 Helicopter-Borne Morocco Desert Campaign

Stefano Perna [1,2,*], Giovanni Alberti [3], Paolo Berardino [1], Lorenzo Bruzzone [4], Dario Califano [3], Ilaria Catapano [1], Luca Ciofaniello [3], Elena Donini [4], Carmen Esposito [1], Claudia Facchinetti [5], Roberto Formaro [5], Gianluca Gennarelli [1], Christopher Gerekos [4], Riccardo Lanari [1], Francesco Longo [5], Giovanni Ludeno [1], Mauro Mariotti d'Alessandro [6], Antonio Natale [1], Carlo Noviello [1], Gianfranco Palmese [3], Claudio Papa [3], Giulia Pica [3], Fabio Rocca [6], Giuseppe Salzillo [3], Francesco Soldovieri [1], Stefano Tebaldini [6] and Sanchari Thakur [4]

[1] Institute for Remote Sensing of Environment (IREA), National Research Council (CNR), 80124 Napoli, Italy
[2] Department of Engineering (DI), Università degli Studi di Napoli "Parthenope", 80143 Napoli, Italy
[3] CO.RI.S.T.A., COnsorzio di RIcerca su Sistemi di Telesensori Avanzati, 80143 Napoli, Italy
[4] Department of Information Engineering and Computer Science, University of Trento, 38123 Trento, Italy
[5] Agenzia Spaziale Italiana, Via del Politecnico, 00133 Roma, Italy
[6] Department of Electronics, Information, and Bioengineering (DEIB), Politecnico di Milano, 20133 Milano, Italy
* Correspondence: stefano.perna@uniparthenope.it; Tel.: +39-081-762-0632

Received: 21 June 2019; Accepted: 5 August 2019; Published: 8 August 2019

Abstract: This work is aimed at showing the present capabilities and future potentialities of an imaging radar system that can be mounted onboard flexible aerial platforms, such as helicopters or small airplanes, and may operate in the UHF and VHF frequency bands as Sounder and as Synthetic Aperture Radar (SAR). More specifically, the Sounder operates at 165 MHz, whereas the SAR may operate either at 450 MHz or at 860 MHz. In the work, we present the first results relevant to a set of Sounder and SAR data collected by the radar during a helicopter-borne campaign conducted in 2018 over a desert area in Erfoud, Morocco, just after the conclusion of a system upgrading procedure. In particular, a first analysis of the focusing capabilities of the Sounder mode and of the polarimetric and interferometric capabilities of the SAR mode is conducted. The overall system, originally developed by CO.RI.S.T.A. according to a ASI funding set up in 2010, has been upgraded in the frame of a contract signed in 2015 between ASI and different private and public Italian Research Institutes and Universities, namely CO.RI.S.T.A., IREA-CNR, Politecnico di Milano and University of Trento.

Keywords: Synthetic Aperture Radar (SAR); Airborne SAR; Sounder; P-Band; helicopter-borne radar; UHF and VHF bands

---

1. Introduction

Radar imaging [1] represents a powerful tool in several applications, such as security, surveillance, and environmental monitoring, with particular emphasis on disasters and crisis management [2–4]. In this frame, radar systems mounted onboard aerial platforms, such as airplanes [5–8], helicopters and drones [9], are gaining increasing interest due to their features allowing to overcome several limitations of radar systems mounted onboard terrestrial vehicles (carts, cars, ground track rails) [10,11] or spaceborne platforms [12–14].

More specifically, if compared to the terrestrial imaging radar systems, such as the conventional Ground Penetrating Radar (GPR) [10] or the Ground Based (GB) Synthetic Aperture Radar (SAR) [11], the aerial imaging radar systems are flexible and able to ensure a wider spatial coverage. Furthermore, in critical scenarios, as the ones induced for instance by natural disasters (Earthquakes, volcanoes' eruptions, flooding, rapid landslides), ungentle climatic conditions (cryosphere monitoring) or anthropic actions (ordnance deployment), they allow safe monitoring of areas that would be difficult to reach (if not unreachable) through terrestrial systems.

On the other side, in contrast to the satellite platforms [12–14], the aerial ones allow to timely reach the area of interest, to fly practically along any direction, and to keep very short the so called revisiting time, that is, the time interval elapsing between subsequent observations of the same area.

In addition, the aerial radar systems are relatively cheap, thus offering the appealing opportunity of assessing the potentialities of novel technologies and/or measurement modalities, before these are made operative on expensive spaceborne missions.

In this regard, the increasing interest registered in the last years toward aerial radar systems operating at low frequencies has been in some extent driven by the ongoing or future spaceborne missions. Indeed, the exploitation of GPR systems for the ongoing planetary exploration missions [15–19] has certainly taken benefit from the results obtained through the several experimental campaigns carried out in the last years with aerial penetrating radar (usually named Sounder) systems operating in the HF, UHF and VHF bands [20–33]. For the same reason, in anticipation of the forthcoming spaceborne ESA-Biomass mission [34], increasing of the number of aerial P-Band SAR missions is registered in the last years [35–46].

In this context, the Italian Space Agency (ASI) has recently funded the development of an aerial multi-mode pulsed imaging radar system operating in a multi-frequency modality in the UHF and VHF bands. In particular, the system is able to work either as Sounder or as full polarimetric SAR. The development of this system is aimed at making available to the Italian community of researchers and end-users an aerial imaging radar system with several attractive features. First, it grows the relatively small family of available aerial SAR [35–58] and Sounder [23,24,26,59] systems operating in the UHF and VHF bands. Secondly, the system allows to easily collect over the same area radar data characterized by different scattering mechanisms (since it may work either as SAR or as Sounder), at different carrier frequencies and (for the SAR modes) with a diversity in the polarization.

The overall system has been originally developed by CO.RI.S.T.A. [47] according to an ASI funding set up in 2010. After, the first version of the system has been upgraded, again by CO.RI.S.T.A., in the frame of a contract signed in 2015 between ASI, CO.RI.S.T.A. and different public Italian Research Institutes and Universities, namely IREA-CNR, Politecnico di Milano and University of Trento, which have been entrusted with the processing of the data acquired by the radar. In particular, they have designed ad-hoc strategies to process the both Sounder and SAR data. These strategies exploit the navigation information acquired during the flight and adopt model-based procedures, which are the microwave tomographic imaging for the Sounder and the back-projection procedure for the SAR.

With the aim of obtaining a first assessment of the performances of this radar system, a helicopter-borne campaign has been conducted in 2018 over a desert area in Erfoud, Morocco. During the campaign, several tracks have been flown exploiting all the three different operational modes of the system. In this work, we present first results relevant to a small subset picked up from the huge dataset collected by the system during this campaign.

The work is organized as follows. In Section 2 we provide a brief description of the system. The acquisition campaign is described in Section 3. The processing chain applied to the Sounder data is illustrated in Section 4 along with some first results. Similarly, the SAR data processing chain is described in Section 5 along with the corresponding first results. Section 6 is devoted to the concluding remarks.

2. System Description

The radar system exploits the pulsed radar technology and can operate at different carrier frequencies as Sounder and Synthetic Aperture Radar (SAR). More specifically, the Sounder operates at

165 MHz, whereas the SAR may operate either at 450 MHz (SAR-Low mode) or at 860 MHz (SAR-High mode). Summing up, three operational modes are enabled, namely, Sounder, SAR-Low and SAR-High, each working at a different frequency.

The bandwidth of the transmitted (chirp) pulses is 40 MHz, leading to a (slant) range resolution of 3.75 m (in air). Moreover, in the SAR-High mode, through the stepped chirp technology [60], an overall bandwidth of 80 MHz can be reached, that is, a (slant) range resolution of 1.87 m. The SAR system is full polarimetric thanks to two separate receiving channels.

The overall system basically consists of a radar module along with three different antennas, one for each operational mode.

The radar module consists of three main blocks, namely, the Radar Digital Unit (RDU), the Radio Frequency Unit (RFU) and the Power Supply Unit (PSU). In particular, the RDU is fully programmable and allows the parameters setting, the timing generation and the data handling. It also includes the Analog to Digital Converter (ADC) and the data storage unit. The RFU embeds the frequency generation unit (which generates all the synchronization and radio frequency signals) and the chirp generator unit (which generates the low frequency modulated chirp signal by means of the digital direct synthesis technology). RFU also includes the high power amplification unit, which is based on the solid state technology, and an antenna front-end that allows the correct switching among the transmitted and received signals and among the different polarizations of the SAR mode. The PSU provides the power supply to whole system by an external 28 V DC voltage.

The system is completely stand-alone: the power supply connector is the only electrical interface. Most radar modules are shared by the three different operational modes of the system. In particular, base band signal generation, base band data sampling and data handling are common to the Sounder and the SAR modes.

As noted above, different antennas are used for the three different operational modes of the system. In particular, for the Sounder mode, a low-gain log-periodic antenna is deployed; the antenna points to the nadir and radiates the main beam with Half Power Beam Width (HPBW) [61] of 68° and 50° in the range and azimuth directions, respectively, and with a (maximum) gain [61] of 7dBi[1]. Exploitation of the overall azimuth aperture thus can lead to an azimuth resolution of about 1m [62,63]. Two different side-looking arrays of micro-strip antennas are instead employed for the two different SAR modes [64]. In particular, these two SAR antennas have the same shape and size, to allow an easier swap when they cannot be installed simultaneously on the airplane/helicopter. More specifically, the array used for the SAR-Low mode consists of 4 patches deployed along the azimuth direction: its (maximum) gain is 11 dBi and its HPBWs are 75° and 20° in the range and azimuth directions, respectively. Exploitation of the overall azimuth aperture thus leads to an azimuth resolution of about 1m [62,63]. For the SAR-High mode (which operates at a carrier wavelength scaled about of a factor two respect to the SAR-Low mode), a planar array antenna of 8×2 patches is deployed, with the same overall dimensions of the array used for the SAR-Low mode. The corresponding HPBWs are thus reduced of a factor two (in both directions) with respect to the SAR-Low mode. This allows increasing the (maximum) antenna gain of 6 dBi, retaining practically the same achievable azimuth resolution obtainable with the SAR-Low mode. Both SAR antennas are dual polarized, to enable the full polarimetric capability of the SAR modes.

Currently, during the flight the user can select manually one of the three operational modes of the system. At present, the different operational modes cannot be used simultaneously; development of a synchronization solution aimed at circumventing this limitation is however straightforward and is matter of current activities.

In order to achieve accurate flight information, necessary for the reliable processing of both Sounder and SAR data, the radar encompasses a navigation unit consisting of an Inertial Navigation System (INS) that embeds a Global Positioning System (GPS). In particular, the INS is a MTi model of Xsense Technologies B.V., which is an inertial measurement unit with integrated 3D magnetometers (3D compass), with an embedded processor capable of calculating roll, pitch and yaw in real time, as well as outputting calibrated 3D linear acceleration, rate of turn (gyro) and (Earth) magnetic field data. The INS device integrates a GPS receiver and the related information, position, velocity and time (PVT). The navigation unit is directly connected to the radar central unit by means of an USB interface; all the navigation data are synchronized with the radar pulses and embedded in its output data. Furthermore, two additional GPS devices have been added to allow differential processing of the positioning data. The two GPS receivers are based on M8T chip series of U-blox, one installed onboard, close to the SAR antenna, and the other on the ground in the range of some kilometers from the flying radar. During the calibration campaigns of the radar, differential processing of the positioning data, carried out with the RTKLIB library, allowed a centimetric positioning accuracy.

The overall system is easy to be transported and installed onboard relatively small airplanes or helicopters. Indeed, the radar module is stowed in a rack, which is quite compact and with dimensions of 50 cm × 50 cm × 65 cm, for a weight of about 30 kg.

During the campaign, the radar system was installed on an Eurocopter AS-350 series helicopter (see Figure 1a). In particular, the radar electronics rack was accommodated in the helicopter cabin, in place of the two rear passenger seats (see Figure 1b). The sounder antenna is installed on the helicopter nose (see Figure 1c) through the mechanical framework normally used for handling the rear-view mirror; moreover, it is fixed to the front glass and to the fuselage by means of a standard adapter. The SAR antennas are installed through a certified framework normally used for side-looking cameras and by using a honeycomb panel and some brackets (see Figure 1d, where just one SAR antenna is shown).

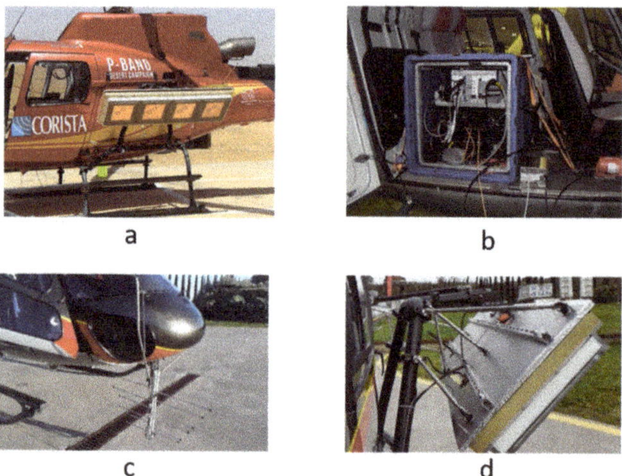

Figure 1. (**a**) The system ready to flight, (**b**) radar electronics rack inside the cabin, (**c**) Sounder antenna, (**d**) SAR antenna.

Table 1 reports the main electrical, mechanical and geometrical parameters of the system.

Table 1. System parameters.

Parameters	Sounder	SAR-Low	SAR-High
Carrier frequency	165 MHz	450 MHz	860 MHz
Bandwidth	40 MHz	40 MHz	80 MHz
Frequency steps	1	1	2–3
PRF	500 Hz	1000 Hz	2000–3000 Hz
Pulse width	2 μs	2 μs	2 μs
Mode	Pulsed	Pulsed	Pulsed
Antenna type	Log Periodic	array of microstrip antennas	array of microstrip antennas
Antenna gain	7 dBi *	11 dBi *	17 dBi *
Antenna dimension	8 × 58 × 107 cm	10 × 45 × 165 cm	10 × 45 × 165 cm
Antenna weight	2 Kg	15 Kg	15 Kg
Nominal elevation pointing	Nadir	45°	45°
Nominal azimuth pointing	Nadir	0°	0°
Range aperture	68°	75°	37°
Azimuth aperture	50°	20°	10°
Range resolution	3.8 m (free sp.)	3.8 m	1.9 m
Azimuth resolution	1 m	1 m	1 m
ADC Sampling frequency	200 MHz		
Peak power	200 W		
Power consumption	500 W		
Rack Weight	30 Kg		
Rack Dimension	50 × 50 × 65 cm		

* dBi measures the gain of an antenna with respect to an isotropic radiator.

3. Campaign Description

The flight campaign was carried out in May 2018 over an arid region located in southeastern Morocco, around the city of Erfoud at the northern edge of the Sahara Desert. Several tracks, covering different test sites, have been flown by exploiting all the three available radar acquisition modes. As observed above (see again Figure 1), the radar system was installed onboard an Eurocopter AS-350 series helicopter, and it was possible to embark onboard simultaneously the Sounder antenna and only one SAR antenna. Exploitation of the two different SAR modes has thus required during the mission the swap of the two different SAR antennas.

In order to minimize the effects of structural vibration noise, during the campaign the INS has been placed inside the helicopter cabin (as close as possible to the SAR antenna) on the floor, by using vibration absorbers.

Figure 2 shows all the tracks flown during the campaign, superimposed to an optical image of the overall test area. In particular, the red, green and blue tracks are relevant to the acquisitions carried out with the Sounder, SAR-Low and SAR-High modes, respectively. As can be seen, the flight tracks were flown basically over four different test sites that in the following are named TS1, TS2, TS3 and TS4, respectively. More specifically, the test site TS1, which is located in proximity of the cities of Errachidia and Erfoud, covers an aquifer located in cretaceous rocks at a depth ranging approximatively between 100 m and 200 m. The test sites TS2 and TS3 are both located in proximity of the city of Rissani. In particular, TS2, which belongs to the great Basin Ziz Rheris, is an almost flat region characterized by alluvial sediments with the presence of aquifers at depth varying approximatively between 20 m and 40 m, whereas TS3 includes the archeological area of Sijilmasa. The test site TS4 is located in proximity of the Erg Chebbi and is characterized by the presence of several dunes and covers an aquifer located at a depth ranging approximatively between 150 m and 200 m.

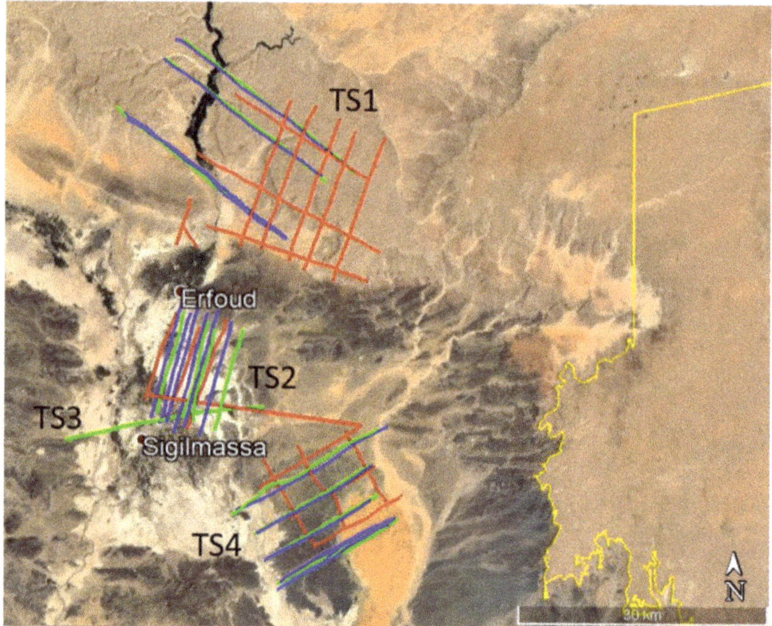

Figure 2. Erfoud area, southeastern Morocco. The red, green and blue lines represent the projections on the ground of the flight tracks relevant to the acquisitions carried out with the Sounder, SAR-Low and SAR-High modes, respectively.

4. Data Processing and Experimental Results: Sounder Mode

Sounder raw data acquired during the campaign have been delivered by CO.RI.S.T.A. and processed by IREA-CNR and University of Trento.

4.1. Data Processing

Like any penetrating radar, the Sounder data collected onboard, usually named raw-data, are arranged in a two-dimensional matrix representing the radar echoes as a function of the fast time and the slow time, which are related to the two-way sensor-to-target distance (range) and the position of the sensor along the flight trajectory (azimuth), respectively. The main rationale of the processing chain applied to focus these raw data is shown in Figure 3 and consists of the cascade of four steps.

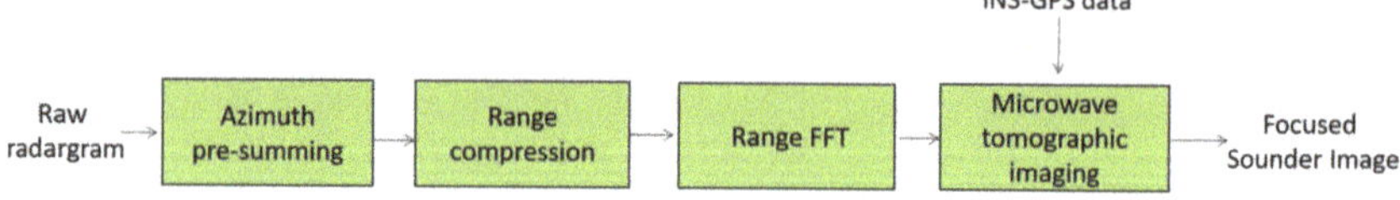

Figure 3. Sounder data processing.

The first step is an azimuth pre-summing aimed at improving the signal-to-noise ratio. In this regard, it is noted that the Sounder PRF (500 Hz, see Table 1) leads to a spectral window significantly larger than the Doppler bandwidth (about 40 Hz) achievable with the available azimuth beam (50°, see again Table 1) and the platform velocity (on the order of 40 m/s). Accordingly, the pre-summing factor can be safely set on the order of 10 or even more. In the cases at hand, a factor of 20 has been applied along with a proper anti-aliasing filtering procedure.

The second processing step (see again Figure 3) is the chirp pulse compression [65], which includes also the application of a spectral windowing procedure (in the case at hand, we use Hamming window) aimed at reducing the level of the secondary lobes of the impulse response relevant to the range-focused image.

Following this step, the data are still not focused along the azimuth direction: they are thus generally characterized by the presence of diffraction hyperbolas that can make very difficult the interpretation of the radar image. The data interpretation is further complicated by the fact that the Sounder acquisition is typically corrupted by the so called motion errors, that is, the deviations of the flown flight track with respect to an ideal rectilinear path. Indeed, the motion errors, if not properly accounted for, produce distortion effects on the final radar image. For instance, in the presence of altitude variations, a flat surface (for instance a lake) is imaged in the Sounder data as a curve specular to the flight path instead of being seen as flat [66]. Application of a proper data processing step aimed at reducing all these effects is thus needed at this stage. In our case, the azimuth focusing procedure, including the compensation of the motion errors, is carried out in the frequency domain. In particular, following the application of a Fast Fourier Transform (FFT) algorithm along the range direction (see again the processing chain of Figure 3), the microwave tomography algorithm described below is applied.

To this aim, let us consider the two-dimensional Sounder acquisition geometry shown in Figure 4. Γ is the flight path that, as observed above, is typically not a straight line, that is, the flight altitude with respect to the ground is variable. $\underline{r}_m$ represents the generic measurement position, that is, the antenna phase center position at a generic azimuth time, say t_m. Let D be the 2D domain of investigation within which we assume the targets of interest are present and let $\underline{r}$ be the position of a generic point in D. The applied tomographic reconstruction approach is based on the following assumptions:

- the scene under investigation is invariant along the direction perpendicular to the plane defined by the normal to the ground and the (nominal) flight direction (2D geometry);
- the radar antenna is located in the far zone with respect to the investigated area. Moreover, the transmitting antenna is modeled as an electrical line source perpendicular to the survey plane;
- the propagation in the soil is negligible because the flight altitude is much greater than the depth of the target of interest. This approximation is acceptable when the goal is to locate shallow targets;
- the measurement configuration is multimonostatic/multifrequency so that, at each measurement point $\underline{r}_m$, the radar illuminates the scene at nadir and collects the field scattered by the targets at the same measurement point $\underline{r}_m$, within the frequency band of the transmitted signal;
- a linearized model of the electromagnetic scattering equations is exploited [67].

According to the above assumptions, the electromagnetic scattering phenomenon is governed by the integral equation [67]:

$$E_s\left(\underline{r}_m, \omega\right) = k_0^2 \iint_D g\left(\underline{r}_m, \underline{r}\right) E_{inc}(\underline{r}) \chi(\underline{r}) d\underline{r} = \mathcal{A}\chi \tag{1}$$

where E_s is the co-polar component of the field scattered by the target (data) and received by the antenna in $\underline{r}_m$ at the frequency ω, k_0 is the propagation constant in free-space, E_{inc} is the incident electric field in D, χ is the so called contrast function, which is different from zero in the target regions and zero elsewhere. Moreover, $\mathcal{A}$ is a linear operator that maps the space of the unknowns in the data space and

$$g\left(\underline{r}_m, \underline{r}\right) \approx \frac{e^{jk|\underline{r} - \underline{r}_m|}}{|\underline{r} - \underline{r}_m|} \tag{2}$$

accounts for the scalar Green's function in free-space [67]. The inverse problem defined by the Equation (1) is ill posed; adoption of a regularization scheme is thus necessary to obtain a stable and robust solution with respect to the data noise [68]. A simple inversion scheme that provides a regularized solution to the inverse problem is that based on the adjoint operator $\mathcal{A}^+$ [68]:

$$\chi(\underline{r}) = \mathcal{A}^+ E_s = \iint \left(g(\underline{r}_m, \underline{r}) E_{inc}(\underline{r}) \right)^* E_s(\underline{r}_m, \omega) d\underline{r}_m d\omega \tag{3}$$

where the symbol '*' denotes the conjugation operator. The module of the contrast function $|\chi(\underline{r})|$, obtained through the Equation (3), represents the focused tomographic image I to be used for the subsequent analysis and interpretation. It is worth stressing that, despite the adjoint inversion scheme defined by Equation (3) is computationally efficient, the computational resources required are in any case high for the imaging problem at hand since large-scale (in terms of probing wavelength) data have to be processed. For this reason, we decided to adopt a shifting zoom approach [69], whose main steps are shown in Figure 5. More specifically, the measurement domain Γ and the survey area D are first partitioned into N partially overlapping subdomains, say Γ_i and D_i, $i = 1, \dots, N$. Then, for each subdomain D_i the reconstruction I_i is obtained by the adjoint inversion. The tomographic image relevant to the whole survey area D is finally obtained by combining the reconstructions I_i relating to the subdomains D_i. Note that splitting the data processing domain in partially overlapping subintervals reduces the computational burden and in particular the memory occupation. In our case, the dimension of the azimuth partitioning has been chosen in the interval [150, 200] m.

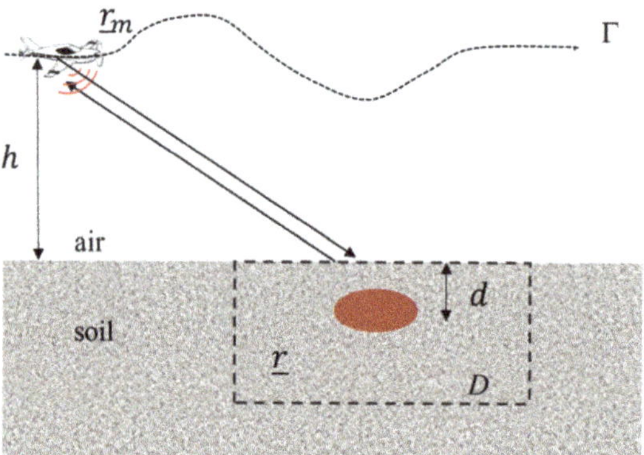

Figure 4. Sounder acquisition geometry.

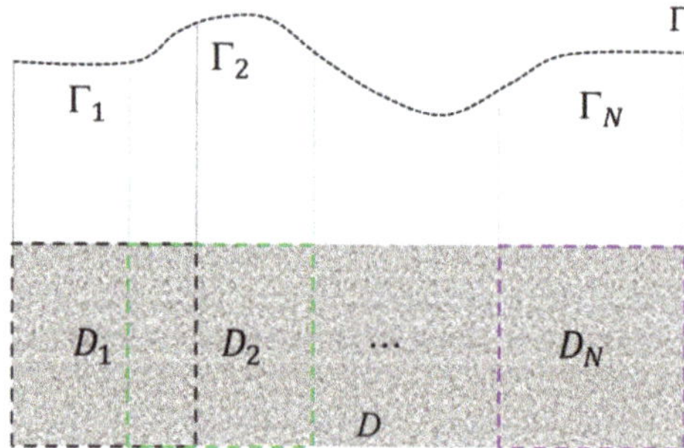

Figure 5. Shifting zoom imaging approach.

Note also that the inversion procedure described above makes it possible to process the data acquired in the presence of arbitrary flight paths, provided that the measurement positions $\underline{r}_m$ are known. In this regard, it is remarked that the accurate position of the measurement points is provided by the embedded INS-GPS device mounted onboard the platform. It is also stressed that, similarly to the SAR case, accounting for the actual flight track leads to an increase in the computational complexity with respect to the focusing procedures that are based upon the assumption of straight-line flight trajectories [70,71].

4.2. First Results

As shown in Figure 2 (red tracks), the sites TS1, TS2 and TS4 were inspected with the Sounder mode. In particular, for each test site, repeated passages of the helicopter have been flown and among the overall acquired dataset, we report in the following some representative reconstruction results.

Figure 6 reports (right panel) the amplitude of the focused Sounder image relevant to a track flown over the site TS1. The left panel reports the Google Earth ortophoto including the observed area. The red line represents the flight track projected on the ground. The focused radar image is arranged in a 2D grid: the vertical axis reports the range coordinate (measured in m); the horizontal axis represents the azimuth coordinate (expressed in latitude). In order to check the accuracy of the focusing procedure in the presence of motion errors, the focused image is compared to the Sounder-to-ground distance (red line) evaluated as the difference between the sensor altitude (provided by the navigation unit mounted onboard the helicopter) and the terrain elevation (provided by the external SRTM Digital Elevation Model). As can be observed in the figure, the red curve is in good agreement with the surface topography. A similar agreement has been observed for all the other datasets, confirming that the instrument, which includes the radar and the navigation unit, is able to accurately reproduce the surface topography variations. Figure 6 also highlights different subsurface returns at apparent depths ranging from a few tens to a few hundred meters. However, their nature is still subject of investigation and interpretation also aided by electromagnetic simulations of surface clutter. Indeed, such simulations account for the true surface tomography provided by the external DEM and can help to avoid interpreting off-nadir surface reflections as nadir subsurface echoes.

Similar considerations hold also for all the tomographic reconstructions achieved over the different test sites inspected with the Sounder mode. Figures 7 and 8 are two representative images (relevant to two tracks flown over the sites TS2 and TS4, respectively), which clearly highlight how the surface clutter due to surface tomography variations and noise can make very challenging the interpretation of Sounder data in realistic scenarios. By the way, the Sounder results shown in the right panels of Figures 6–8, do not allow the detection of subsurface water-bearing structures, which were expected at a depth of 100 m on the basis of a priori information about the geological setting of the site. However, they show several localized subsurface reflections, which are not related to surface clutter according to the electromagnetic scattering simulations that we carried out. The interpretation of these subsurface reflections is currently under investigation and will be subject of a future work.

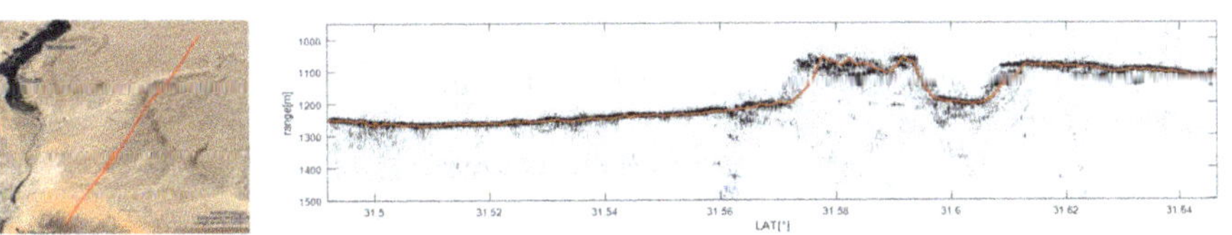

Figure 6. Sounder results relevant to one flight track on TS1. Left panel: Google Earth ortophoto including the observed area. Right panel: Corresponding tomographic reconstruction. Mean flight velocity: 45 m/s. Mean flight altitude: 2132 m. Mean terrain height: 930 m.

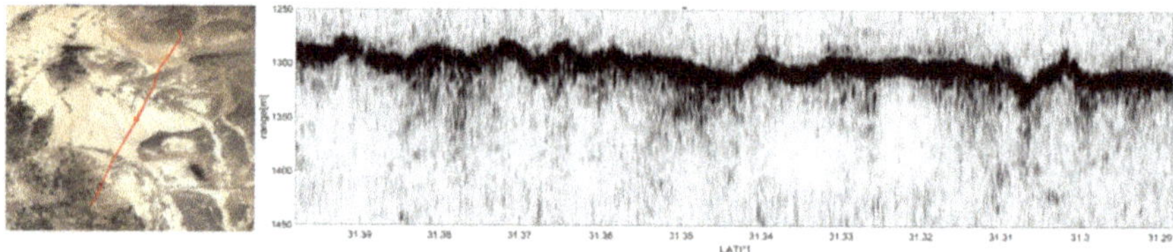

Figure 7. As for Figure 6, but for TS2. Mean flight velocity: 36 m/sec. Mean flight altitude: 2104 m. Mean terrain height: 770 m.

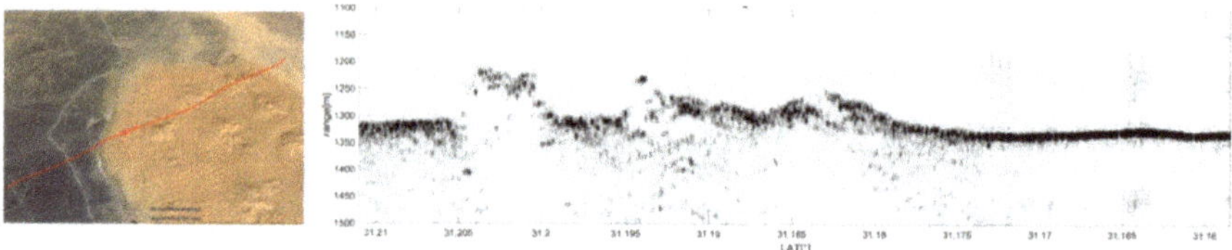

Figure 8. As for Figures 6 and 7, but for TS4. Mean flight velocity: 45 m/sec. Mean flight altitude: 2081 m. Mean terrain height: 745 m.

5. Data Processing and Experimental Results: SAR Mode

SAR raw data acquired during the campaign have been delivered by CO.RI.S.T.A. and processed by IREA-CNR and Politecnico di Milano.

5.1. Data Processing

The SAR data collected onboard, typically named raw-data, are arranged, as in the Sounder case, in a two-dimensional matrix representing the radar echoes as a function of the range and azimuth coordinates.

The main rationale of the processing chain applied to focus these raw data is shown in Figure 9. It basically consists of the cascade of three steps. The first step is the same as that described for the Sounder case, that is, an azimuth pre-summing aimed at improving the signal-to-noise ratio. In this regard, it is noted that the Doppler bandwidth achievable with the available azimuth beam (20° and 10° for the SAR-Low and SAR-High modes, respectively, see Table 1) and the platform velocity (on the order of 40 m/s) is, for both the SAR operational modes, on the order of 40 Hz, that is, significantly smaller than the PRF (which, according to Table 1, becomes 500 Hz when considering each separate polarimetric channel). Accordingly, the pre-summing factor can be safely set on the order of 10 or even more. In the cases at hand, a factor of 8 has been applied.

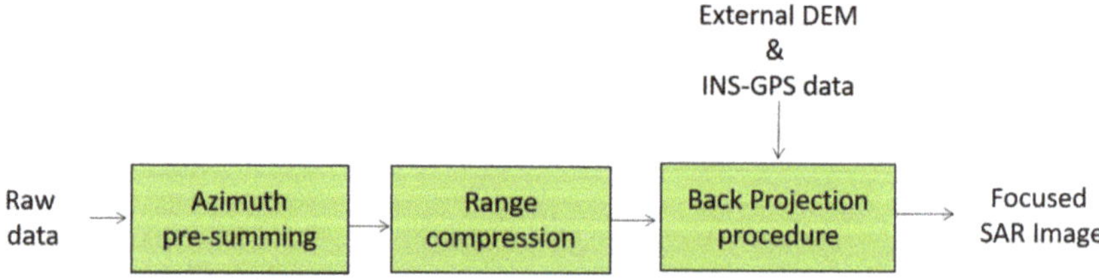

Figure 9. SAR data processing.

Also the second processing step is the same as that of the Sounder processing chain (see again Figure 9): it carries out the range focusing of the data through the chirp pulse compression procedure [62,63], which may include also the application of a spectral windowing aimed at reducing the level of the secondary lobes of the impulse response relevant to the range-focused image. In the case at hand, the Hamming filtering has been applied. Moreover, for the SAR-Low mode data, application of an additional band pass filtering procedure was necessary to cut strong interference contributions well concentrated around the lower frequencies of the bandwidth of the transmitted pulse. In particular, a 25% reduction of the 40 MHz pulse bandwidth has been applied, which has led to a range resolution of about 5 m.

Following this step, azimuth focusing of the data, which includes the correction of the range cell migration effect [63], has been applied. For each target of the considered scene, this procedure basically consists in the compensation of the phase terms that depend on the so-called range history, i.e., the distance between the target and the phase center of the moving radar antenna. In this regard, we note that when the SAR sensor moves along a rectilinear trajectory, as it happens in the spaceborne case (at least theoretically, within a synthetic azimuth length [63]), for all the observed targets the range history is independent of the terrain topography [63]. Differently, when deviations from an ideal straight trajectory occur, as it always happens for aerial platforms (due to the unavoidable presence of atmospheric turbulences), the range history relevant to the observed targets becomes dependent on both the trajectory deviations and the terrain topography [72,73]. Thus, in the latter case, exploitation of external Digital Elevation Models (DEMs) of the observed area, as well as the flight data recorded by the navigation system mounted onboard [72,73], is necessary to precisely compensate the range history of the observed targets. In the last decades a number of accurate and computationally efficient algorithms, which operate in the spectral domain, have been originally devised to focus spaceborne SAR data [63,74] and subsequently modified in order to deal with aerial SAR data as well, through the inclusion of the so-called MOtion COmpensation (MOCO) procedures [72,73]. However, such MOCO procedures involve some approximations [75] necessary to preserve the high computational efficiency of the spectral domain focusing algorithms. In particular, these approximations become unsuitable in the presence of large trajectory deviations or steep topographic variations [73,75] thus impairing the accuracy of the overall focusing procedure [76]. To circumvent these kinds of problems, Back Projection (BP) focusing algorithms [77], which operate in the time domain, are more appropriate, since they do not carry out any approximation, at expenses of a computational efficiency lower than that ensured by the spectral domain focusing algorithms.

In the case at hand, considering the strong track deviations affecting the SAR acquisitions, we applied the BP focusing strategy (see again Figure 9) by exploiting the information provided by the INS-GPS data and the SRTM DEM of the observed area.

5.2. First Results

As shown in Figure 2 (green and blue tracks), all the sites (TS1, TS2, TS3 and TS4) were inspected with the SAR modes. In particular, over the TS1 and TS4 sites, a number of repeat pass interferometric acquisitions were carried out. Accordingly, different full polarimetric as well as interferometric datasets were collected. Among the overall acquired dataset, we report in the following some representative results relevant to the SAR-Low mode data acquired over the TS2 and TS4 sites.

To show the potentialities of this full-polarimetric SAR system, we consider one radar acquisition carried out over the TS2 site.

Application of the processing chain depicted in Figure 9 has led to the generation of four focused Single Look Complex (SLC) data, one for each polarimetric channel. Table 2 reports main data processing parameters. Figure 10 shows, in radar grid (that is, range-azimuth), the amplitude of the corresponding Multi Look Complex (MLC) data relevant to the VV channel. In the right vertical axis of the figure it is specified the (mean) look-angle corresponding to the range coordinate reported in the left vertical axis. The observed scene has an extension of about 9 Km in azimuth and 2.5 Km in range. Figure 11 shows the SAR image of Figure 10 in geographic grid and superimposed to a Google Earth orthophoto. The green line represents the flight track projected on the ground. As can be seen in the orthophoto reported in the left panel, the area exhibits an overall flat topography and it can substantially be considered as a rough bare soil.

Table 2. SAR data processing parameters relevant to the results collected in Figures 10–13.

Parameter	Value	Parameter	Value
Azimuth spacing (SLC)	0.64 m	Range spacing (SLC)	1.5 m
Azimuth resolution (SLC)	1 m	Range resolution (SLC)	5 m
Azimuth spacing (MLC)	0.64 m	Range spacing (MLC)	1.5 m
Azimuth resolution (MLC)	7.5 m	Range resolution (MLC)	12.8 m

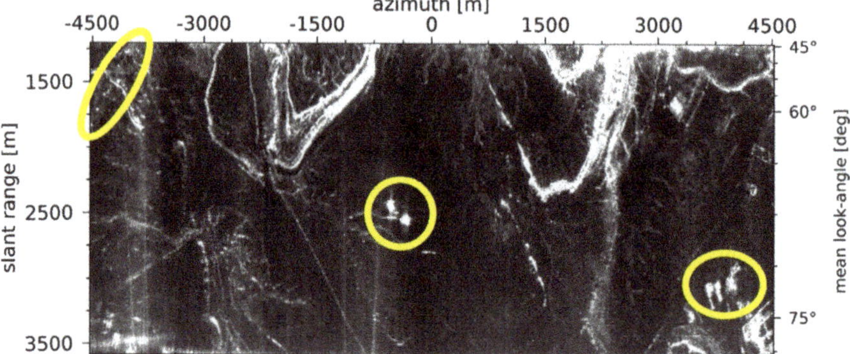

Figure 10. SAR results, relevant to one flight track on TS2. VV channel. SAR-Low mode. Amplitude of the SAR MLC, in radar grid. Mean flight velocity: 40 m/sec. Mean flight altitude: 1664 m. Mean terrain height: 815 m.

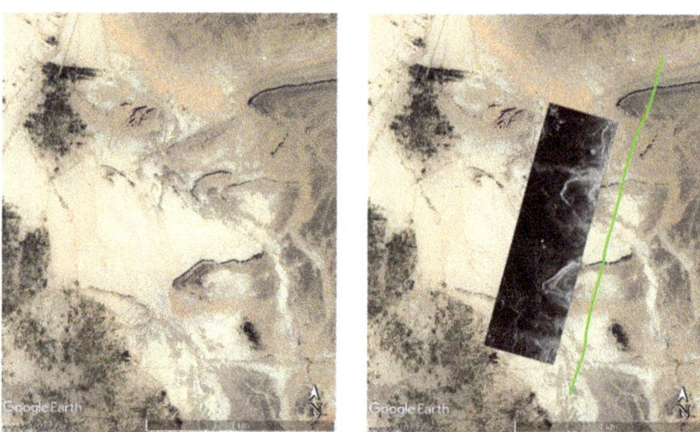

Figure 11. Left panel: Google Earth ortophoto including the observed area. Right panel: As Figure 10, but represented in geographic grid and superimposed to the Google Earth orthophoto of the left panel.

The amplitude and phase of the correlation coefficients between the polarimetric channels are depicted in Figure 12. In particular, see Figure 12a, the amplitude of the co-polar (HH-VV) correlation coefficient assumes quite high values (greater than 0.6) up to a range distance of about 3 Km, corresponding to a mean look-angle of about 70°, while the average phase difference between these channels is very small (approximately $\pi/8$), see Figure 12b. From Figure 12a it can be also observed that the amplitude of the average normalized cross-products between the co-polar (HH and VV) and cross-polar (HV and VH) channels is very low, as it happens when the observed targets exhibit reflection symmetry [78]. These results are thus compatible with the scattering behavior expected in the presence of rough bare soil. The presence of rough bare soil, whose dominant scattering mechanism is the surface scattering contribution [78,79], is confirmed also by the results of the obtained Pauli decomposition [78], which is depicted in Figure 13, where the class of the odd-bounce scattering events (the greenish pixels) is strongly dominant up to distances of about 3 Km. Beyond this range, the transmitted waves impinge the surface at grazing incidence (look angles greater than 70°), thus leading to a non-negligible diffuse scattering contribution (the reddish pixels in Figure 13), and to HH and VV returns which are almost completely uncorrelated, see Figure 12a.

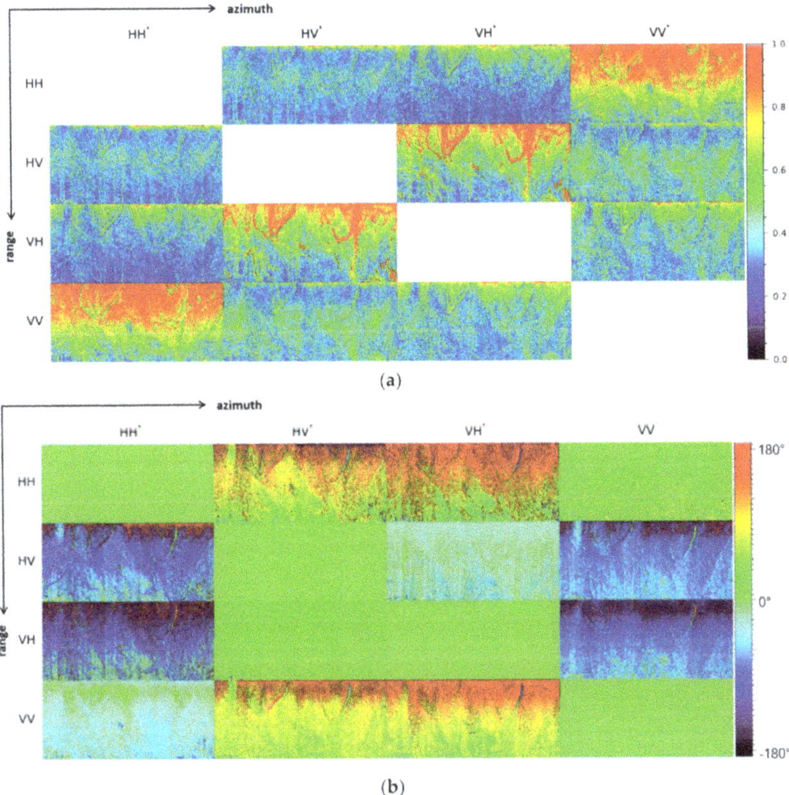

Figure 12. Relevant to the SAR data-set collected by the flight track considered in Figure 10. Correlation coefficients between polarimetric channels: Amplitudes (**a**) and phases (**b**).

Exceptions to the above considerations are represented by some isolated man-made targets which are present in the scene (see, the targets highlighted with the yellow ellipses in Figures 10 and 13). In this case indeed the average phase difference between the HH and VV channels is nearly π, see Figure 12b, and the even-bounce component (the blueish pixels in Figure 13) represents the dominant contribution to the scattering mechanism.

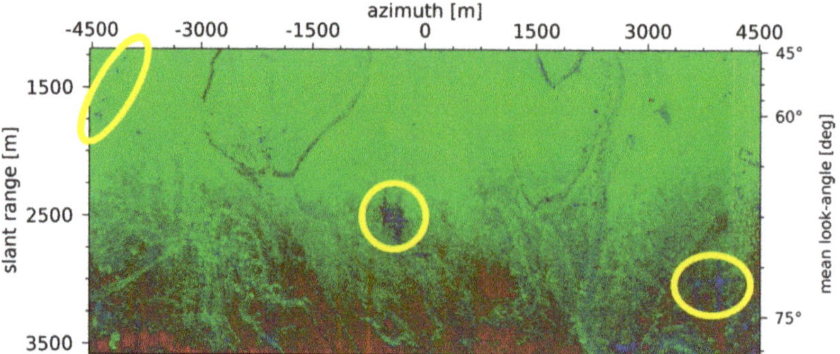

Figure 13. SAR results relevant to the data-set of Figure 12. Pauli decomposition of the considered scene (red: Volume, green: Surface, blue: Double-bounce).

To show the interferometric potentialities of this SAR system, we consider three repeat pass acquisitions carried out over the TS4 site. Few minutes elapsed between the different acquisitions.

Application of the processing chain depicted in Figure 9 has led to the generation of twelve focused SLC, one for each track and polarimetric channel. Table 3 reports main data processing parameters. As an example, Figure 14 shows, in the ground range-azimuth grid, the amplitude of the corresponding MLC relevant to the HH channel. The observed scene has an extension of about 14 Km in azimuth and 3 Km in (ground) range. Figure 15 shows the SAR image of the top panel of Figure 14 in geographic grid and superimposed to a Google Earth orthophoto.

For all the interferometric data pairs achievable with the different polarimetric channels, we have carried out a first interferometric analysis, by using the processing parameters listed in Table 4. In particular, Figure 16 shows the interferometric coherence relevant to the VV channels, which turned out to be on average the highest one with respect to that achievable with the other combinations of polarimetric channels. As can be seen, in most of the observed area the obtained coherence is quite good. Coherence losses are however observed in near range and in some well confined azimuth strips, due to the spatial decorrelation effects induced by the large spatial baselines generated by the severe track deviations of the helicopter during the radar acquisitions. This is confirmed by the height of ambiguity maps reported in Figure 17 and relevant to the same data pairs considered in Figure 16. As can be seen, in the low coherence regions of Figure 16 the height of ambiguity reported in Figure 17 reaches very low (absolute) values. Note in particular, that the color scale in Figure 17 is set between (−15 m, 15 m) to highlight that trajectories are crossing, causing the interferometric baseline to go to 0. Accordingly, at some locations the height of ambiguity becomes infinite (hence all the saturated areas), and then returns in the interval (−15, 15) with opposite sign.

Table 3. SAR data processing parameters relevant to the results collected in Figures 14 and 15.

Parameter	Value	Parameter	Value
Azimuth spacing (SLC)	0.7 m	Ground range spacing (SLC)	5 m
Azimuth resolution (SLC)	1 m	Ground range resolution (SLC)	6.25 m
Azimuth spacing (MLC)	0.7 m	Ground range spacing (MLC)	5 m
Azimuth resolution (MLC)	7 m	Ground range resolution (MLC)	6.25 m

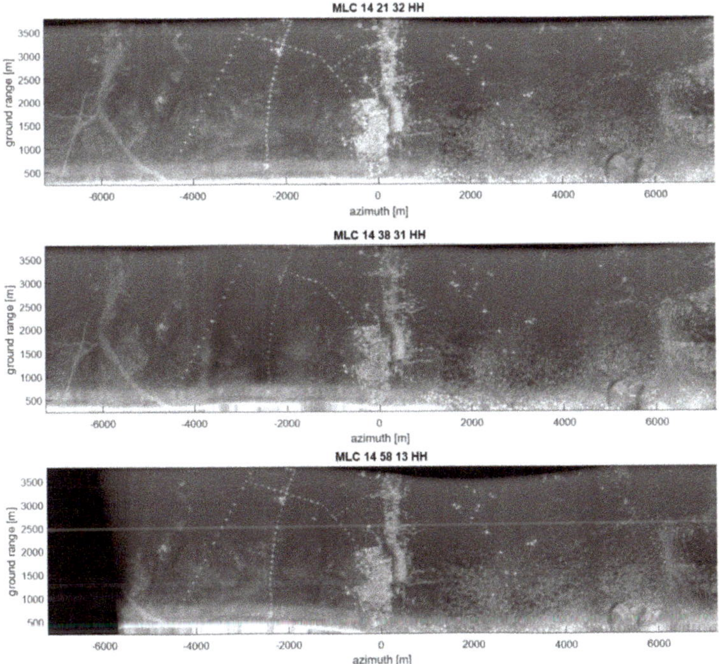

Figure 14. SAR results relevant to three repeat pass flight tracks on TS4. HH channel, SAR-Low mode: Amplitude of the SAR MLCs in ground range-azimuth grid. Acquisition date: 15 May 2018. Mean flight velocity: 50 m/sec. Mean flight altitude: 1600 m. Mean terrain height: 730 m. From the top to the bottom panel the acquisition start time is: 14:21:32, 14:38:31, 14:58:13.

Figure 15. As the top panel of Figure 10 but represented in geographic grid and superimposed to a Google Earth orthophoto.

Table 4. SAR data processing parameters relevant to the InSAR results collected in Figures 16–18.

Parameter	Value	Parameter	Value
Azimuth spacing	4 m	Ground range spacing	6.25 m
Azimuth resolution	15 m	Ground range resolution	15 m

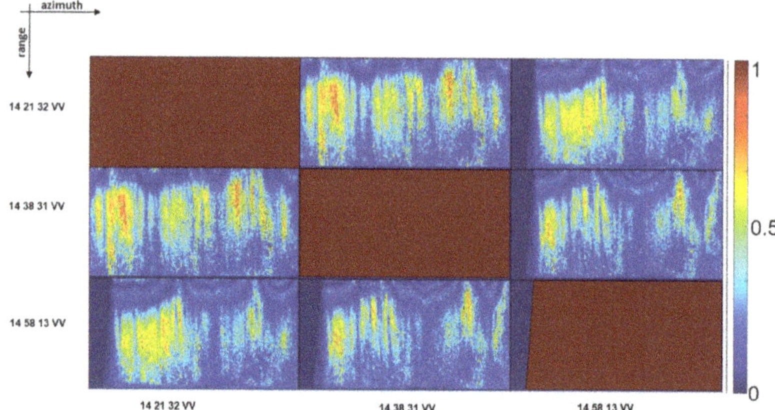

Figure 16. Relevant to the repeat pass interferometric SAR data-set collected by the flight tracks considered in Figure 14. Coherence maps relevant to the VV channels.

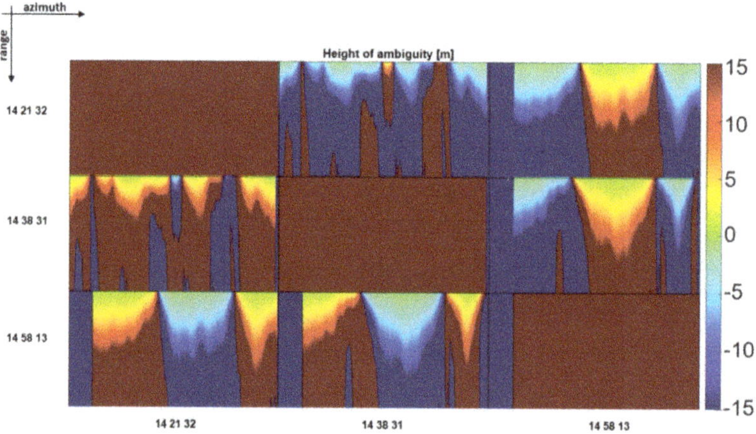

Figure 17. Height of ambiguity maps relevant to the data pairs considered in Figure 16.

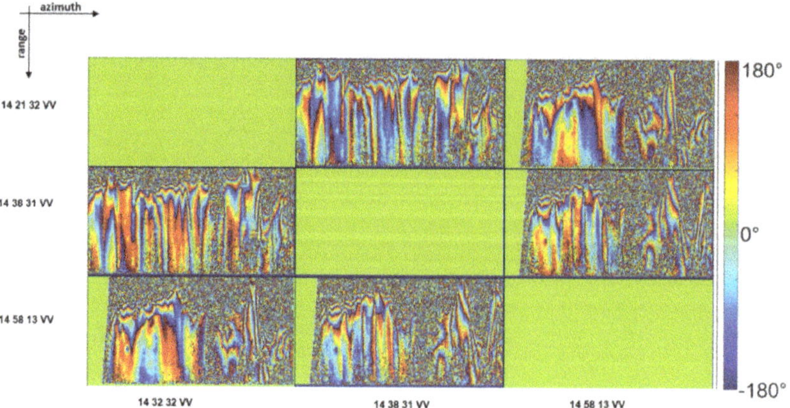

Figure 18. Interferograms relevant to the data pairs considered in Figures 16 and 17.

Note also that in some confined range strips (see for instance the central azimuth portion of the 14:58:13–14:38:31 and 14:21:32–14:58:13 interferograms) the coherence is low from near to far range although the (absolute) value of the height of ambiguity is very high (greater than 15 m) between mid and far range, approximately. This is due to the strong attitude variations of the helicopter during the acquisitions. In some portions of the flown tracks it is indeed likely that these effects were significantly different for the different acquisitions, thus reducing the overlapping between the Doppler spectra of the different interferometric channels.

By the way, in the high coherence regions, which represent the large part of the observed area, interferometric fringes are well visible, as it can be seen in Figure 18, where the interferograms relevant to the same data pairs considered in Figure 16 are shown.

6. Conclusions and Further Developments

In this work, we have presented a recently developed aerial imaging radar system operating in the UHF and VHF bands as Sounder and as full polarimetric Synthetic Aperture Radar (SAR). More specifically, three operational modes are possible: Sounder, SAR-Low and SAR-High, each working at a different frequency.

To obtain a first evaluation of the potentialities of the system, a helicopter-borne campaign has been conducted in May 2018 over an arid region located in southeastern Morocco, around the city of Erfoud at the northern edge of the Sahara Desert. Several tracks, covering four different test sites (denoted as TS1, TS2, TS3 and TS4), have been flown by exploiting all the three available radar acquisition modes. From the huge dataset collected during the campaign, we have processed a small subset and presented first results.

In particular, to show the potentialities of the Sounder system, we have presented some representative results relevant to the TS1, TS2 and TS4 sites. To focus the data, we have applied a tomographic reconstruction approach capable to deal with the actually flown flight tracks by exploiting the information provided by the INS-GPS system mounted onboard the helicopter. To check the accuracy of the obtained results, the focused images have been compared to the Sounder-to-ground distance evaluated by exploiting the INS-GPS system and the external SRTM DEM. A good agreement between expected and obtained results has been achieved.

To show the potentialities of the SAR system, we have shown some representative results relevant to the SAR-Low mode data acquired over the TS2 and TS4 sites. More specifically, to focus the data, we have applied a Back Projection approach operating in time domain and capable of exploiting the information provided by the INS-GPS system mounted onboard the helicopter and the external SRTM DEM of the observed area. To show the potentialities of the system related to its full-polarimetric capability, we have considered one radar acquisition carried out over the TS2 site and shown the correlation coefficients between the polarimetric channels as well as the obtained Pauli decomposition. It has been shown that these polarimetric products well match the scattering behavior expected for the observed area (which can substantially be considered as a rough bare soil in the presence of a limited number of well confined man-made targets). To show the interferometric potentialities of the system, we have considered three repeat pass acquisitions carried out over the TS4 site and shown the obtained interferograms along with the corresponding coherence maps. In particular, it has been shown that the obtained interferometric coherence is quite good, but for the near range areas and some well confined azimuth strips, due to the spatial decorrelation effects induced by the large spatial baselines generated by the severe track deviations of the helicopter during the radar acquisitions. By the way, in the high coherence regions, which represent the large part of the observed area, interferometric fringes are well visible.

Summing up, first results relevant to both the Sounder and the SAR modes are promising. Of course, further activities aimed at assessing the full capabilities of the system are planned for the next months.

In particular, full exploitation of the available SAR polarimetric channels in order to obtain added value products such as soil moisture as well as surface roughness maps [80] is matter of current investigation. Beside, acquisition of massive repeat pass interferometric data-sets is planned for the very near future, in order to fully exploit the capabilities of the UHF and VHF bands for the retrieval of the terrain topography below dense forests [81] or the subsurface structure over areas covered by snow/ice [42] through advanced tomographic SAR processing techniques. Moreover, deep interpretation of the nature of subsurface returns visible in the focused Sounder images is subject of investigation and interpretation also aided by electromagnetic simulations of surface clutter. In the meantime, processing of the entire dataset acquired during the Morocco campaign is matter of current study and future work.

Author Contributions: Conceptualization, S.P., F.S., L.B., S.T., G.A., D.C., L.C., G.P., C.P. and G.S.; Methodology, S.P., F.S., L.B., S.T. and G.A.; Software and Validation, P.B., I.C., C.E., G.G., G.L., A.N., C.N., E.D., C.G., S.T. and M.M.d.; Formal Analysis, S.P., F.S., L.B., S.T. and G.A.; Investigation, S.P., F.S., L.B., S.T., G.A., P.B., I.C., C.E., G.G., G.L., A.N., C.N., E.D.., C.G., S.T. and M.M.d.; Data Curation, D.C., L.C., G.P., C.P., G.S., P.B., I.C., C.E., G.G., G.L., A.N., C.N., E.D., C.G., S.T. and M.M.d.; Writing-Original Draft Preparation, S.P.; Supervision, S.P., F.S., L.B., S.T., G.A., R.L. and F.R.; Project Administration, C.F., R.F., F.L. and G.P.; Funding Acquisition, R.L., L.B., F.R. and G.A.

Funding: The work has been partly funded by the Italian Space Agency under the contract N. 2015-029-I.0 "Evoluzione tecnologica e sperimentazione, tramite piattaforma aerea, di un sensore Radar nelle bande VHF e UHF (frequenze inferiori ad 1 GHz)".

Acknowledgments: The authors thank the anonymous reviewers for their comments and suggestions.

Conflicts of Interest: The authors declare no conflict of interest.

References

1. Tsang, L.; Kong, J.A.; Shin, R.T. *Theory of Microwave Remote Sensing*, 1st ed.; Wiley-Interscience: Hoboken, NJ, USA, 1985.

2. Massonnet, D.; Briole, P.; Arnaud, A. Deflation of Mount Etna monitored by spaceborne radar interferometry. *Nature* **1995**, *375*, 567–570. [CrossRef]

3. Lanari, R.; Berardino, P.; Bonano, M.; Casu, F.; Manconi, A.; Manunta, M.; Manzo, M.; Pepe, A.; Pepe, S.; Sansosti, E.; et al. Surface displacements associated with the L'Aquila 2009 Mw 6.3 earthquake (central Italy): New evidence from SBAS-DInSAR time series analysis. *Geophys. Res. Lett.* **2010**, *37*, L20309. [CrossRef]

4. Lavecchia, G.; Castaldo, R.; de Nardis, R.; de Novellis, V.; Ferrarini, F.; Pepe, S.; Brozzetti, F.; Solaro, G.; Cirillo, D.; Bonano, M.; et al. Ground deformation and source geometry of the 24 August 2016 Amatrice earthquake (Central Italy) investigated through analytical and numerical modeling of DInSAR measurements and structural-geological data. *Geophys. Res. Lett.* **2016**, *43*, 12–389. [CrossRef]

5. Kim, D.-J.; Hensley, S.; Yun, S.-H.; Neumann, M. Detection of Durable and Permanent Changes in Urban Areas Using Multitemporal Polarimetric UAVSAR Data. *IEEE Geosci. Remote Sens. Lett.* **2016**, *13*, 267–271. [CrossRef]

6. Horn, R.; Nottensteiner, A.; Scheiber, R. F-SAR—DLR's advanced airborne SAR system onboard DO228. In Proceedings of the 7th European Conference on Synthetic Aperture Radar, Friedrichshafen, Germany, 2–5 June 2008.

7. Baqué, R.; Bonin, G.; du Plessis, O.R. The airborne SAR-system: SETHI airborne microwave remote sensing imaging system. In Proceedings of the 7th European Conference on Synthetic Aperture Radar, Friedrichshafen, Germany, 2–5 June 2008.

8. Perna, S.; Esposito, C.; Amaral, T.; Berardino, P.; Jackson, G.; Moreira, J.; Pauciullo, A.; Vaz Junior, E.; Wimmer, C.; Lanari, R. The InSAeS4 airborne X-band interferometric SAR system: A first assesment on its imaging and topographic mapping capabilities. *Remote Sens.* **2016**, *8*, 40. [CrossRef]

9. Aguasca, A.; Acevo-Herrera, R.; Broquetas, A.; Mallorqui, J.J.; Fabregas, X. ARBRES: Light-Weight CW/FM SAR Sensors for Small UAVs. *Sensors* **2013**, *13*, 3204–3216. [CrossRef] [PubMed]

10. Daniels, D.J. *Ground Penetrating Radar*; John Wiley & Sons: Hoboken, NJ, USA, 2005.

11. Tarchi, D.; Casagli, N.; Fanti, R.; Leva, D.D.; Luzi, G.; Pasuto, A.; Pieraccini, M.; Silvano, S. Landslide monitoring by using ground-based SAR interferometry: An example of application to the Tessina landslide in Italy. *Eng. Geol.* **2003**, *68*, 15–30. [CrossRef]

12. Virelli, M.; Coletta, A.; Battagliere, M.L. ASI COSMO-SkyMed: Mission Overview and Data Exploitation. *IEEE Geosci. Remote Sens. Mag.* **2014**, *2*, 64–66. [CrossRef]

13. Krieger, G.; Moreira, A.; Fiedler, H.; Hajnsek, I.; Werner, M.; Younis, M.; Zink, M. TanDEM-X: A Satellite Formation for High-Resolution SAR Interferometry. *IEEE Trans. Geosci. Remote Sens.* **2007**, *45*, 3317–3341. [CrossRef]

14. Torres, R.; Snoeij, P.; Geudtner, D.; Bibby, D.; Davidson, M.; Attema, E.; Potin, P.; Rommen, B.; Floury, N.; Brown, M.; et al. GMES Sentinel-1 Mission. *Remote Sens. Environ.* **2012**, *120*, 9–24. [CrossRef]

15. Seu, R.; Biccari, D.; Orosei, R.; Lorenzoni, L.V.; Phillips, R.J.; Marinangeli, L.; Picardi, G.; Masdea, A.; Zampolini, E. SHARAD: The MRO 2005 shallow radar. *Plane. Space Sci.* **2004**, *52*, 157–166. [CrossRef]

16. Croci, R.; Seu, R.; Flamini, E.; Russo, E. The SHAllow RADar (SHARAD) onboard the NASA MRO mission. *Proc. IEEE* **2011**, *99*, 794–807. [CrossRef]

17. Picardi, G.; Plaut, J.J.; Biccari, D.; Bombaci, O.; Calabrese, D.; Cartacci, M.; Cicchetti, A.; Clifford, S.M.; Edenhofer, P.; Farrell, W.M.; et al. Radar soundings of the subsurface of Mars. *Science* **2005**, *310*, 1925–1928. [CrossRef]

18. Jordan, R.; Picardi, G.; Plaut, J.; Wheeler, K.; Kirchner, D.; Safaeinili, A.; Johnson, W.; Seu, R.; Calabrese, D.; Zampolini, E.; et al. The Mars express MARSIS sounder instrument. *Plane. Space Sci.* **2009**, *57*, 1975–1986. [CrossRef]

19. Ono, T.; Kumamoto, A.; Nakagawa, H.; Yamaguchi, Y.; Oshigami, S.; Yamaji, A.; Kobayashi, T.; Kasahara, Y.; Oya, H. Lunar radar sounder observations of subsurface layers under the nearside maria of the Moon. *Science* **2009**, *323*, 909–912. [CrossRef]

20. Cook, J.C. Proposed monocycle-pulse very-high-frequency radar for airborne ice and snow measurement. *Trans. Am. Inst. Electr. Eng. Part I Commun. Electron.* **1960**, *79*, 588–594.

21. Bailey, J.; Evans, S.; Robin, G.D.Q. Radio echo sounding of polar ice sheets. *Nature* **1964**, *204*, 420–421. [CrossRef]

22. Hodge, S.M.; Wright, D.; Bradley, J.; Jacobel, R.; Skou, N.; Vaughn, B. Determination of the surface and bed topography in central Greenland. *J. Glaciol.* **1990**, *36*, 17–30. [CrossRef]

23. Gogineni, S.; Chuah, T.; Allen, C.; Jezek, K.; Moore, R.K. An improved coherent radar depth sounder. *J. Glaciol.* **1998**, *44*, 659–669. [CrossRef]

24. Rodriguez-Morales, F.; Gogineni, S.; Leuschen, C.J.; Paden, J.D.; Li, J.; Lewis, C.C.; Panzer, B.; Alvestegui, D.G.-G.; Patel, A.; Byers, K.; et al. Advanced multifrequency radar instrumentation for polar research. *IEEE Trans. Geosci. Remote Sens.* **2014**, *52*, 2824–2842. [CrossRef]

25. Li, J.; Paden, J.; Leuschen, C.; Rodriguez-Morales, F.; Hale, R.D.; Arnold, E.J.; Crowe, R.; Gomez-Garcia, D.; Gogineni, P. High-altitude radar measurements of ice thickness over the Antarctic and Greenland ice sheets as a part of operation ice bridge. *IEEE Trans. Geosci. Remote Sens.* **2013**, *50*, 742–754. [CrossRef]

26. Hélière, F.; Lin, C.; Corr, H.; Vaughan, D. Radio echo sounding of Pine Island Glacier, West Antarctica: Aperture synthesis processing and analysis of feasibility from space. *IEEE Trans. Geosci. Remote Sens.* **2007**, *45*, 2573–2582. [CrossRef]

27. Kristensen, D.J.; Krozer, S.S.; Hernández, V.; Vidkjær, C.C.; Kusk, J.; Christensen, E.L. ESA's polarimetric airborne radar ice sounder (POLARIS): Design and first results. *IET Radar Sonar Navig.* **2010**, *4*, 488–496.

28. Heggy, E.; Rosen, P.A.; Beatty, R.; Freeman, T.; Gim, Y. Orbiting Arid Subsurface and Ice Sheet Sounder (OASIS): Exploring desert aquifers and polar ice sheets and their role in current and paleo-climate evolution. In Proceedings of the IEEE International Geoscience and Remote Sensing Symposium, Melbourne, Australia, 21–26 July 2013.

29. Bradford, J.H.; Dickins, D.F.; Brandvik, P.J. Assessing the potential to detect oil spills in and under snow using airborne ground-penetrating radar. *Geophysics* **2010**, *75*, G1–G12. [CrossRef]

30. Rignot, E.; Mouginot, J.; Larsen, C.F.; Gim, Y.; Kirchner, D. Low-frequency radar sounding of temperate ice masses in Southern Alaska. *Geophys. Res. Lett.* **2013**, *40*, 5399–5405. [CrossRef]

31. Mouginot, J.; Rignot, E.; Gim, Y.; Kirchner, D.; Le Meur, F. Low-frequency radar sounding of ice in East Antarctica and southern Greenland. *Ann. Glaciol.* **2014**, *55*, 138–146. [CrossRef]

32. Davis, M.E. *Foliage Penetration Radar—Detection and Characterization of Objects under Trees*; SciTech Publishing: Raleigh, NC, USA, 2011.

33. Moussally, G.; Breiter, K.; Rolig, J. Wide-area landmine survey and detection system. In Proceedings of the 10th International Conference on Ground Penetrating Radar, Delft, The Netherlands, 21–24 June 2004.

34. Le Toan, T.; Quegan, S.; Davidson, M.W.J.; Balzter, H.; Paillou, P.; Papathanassiou, K.; Plummer, S.; Rocca, F.; Saatchi, S.; Shugart, H.; et al. The BIOMASS mission: Mapping global forest biomass to better understand the terrestrial carbon cycle. *Remote Sens. Environ.* **2011**, *115*, 2850–2860. [CrossRef]

35. Neumann, M.; Saatchi, S.S.; Ulander, L.; Franson, J. Assessing Performance of L- and P-Band Polarimetric Interferometric SAR Data in Estimating Boreal Forest Above-Ground Biomass. *IEEE Trans. Geosci. Remote Sens.* **2012**, *50*, 714–726. [CrossRef]

36. Hoekman, D.H.; Quiriones, M.J. Land cover type and biomass classification using AirSAR data for evaluation of monitoring scenarios in the Colombian Amazon. *IEEE Trans. Geosci. Remote Sens.* **2000**, *38*, 685–696. [CrossRef]

37. Garestier, F.; Dubois-Fernandez, P.; Guyon, D.; Le Toan, T. Forest Biophysical Parameter Estimation Using L and P band Polarimetric SAR data. *IEEE Trans. Geosci. Remote Sens.* **2009**, *47*, 3379–3388. [CrossRef]

38. Hajnsek, I.; Kugler, F.; Lee, S.K.; Papathanassiou, K.P. Tropical-Forest-Parameter Estimation by Means of Pol-InSAR: The INDREX-II Campaign. *IEEE Trans. Geosci. Remote Sens.* **2009**, *47*, 481–493. [CrossRef]

39. Rott, H.; Davis, R.E. Multifrequency and polarimetric SAR observations on Alpine glaciers. *Ann. Glaciol.* **1993**, *17*, 98–104. [CrossRef]

40. Sharma, J.J.; Hajnsek, I.; Papathanassiou, K.P.; Moireira, A. Polarimetric Decomposition Over Glacier Ice Using Long-Wavelength Airborne PolSAR. *IEEE Trans. Geosci. Remote Sens.* **2011**, *49*, 519–535. [CrossRef]

41. Tebaldini, S.; Nagler, T.; Rott, H.; Heilig, H. Imaging the Internal Structure of an Alpine Glacier via L-Band Airborne SAR Tomography. *IEEE Trans. Geosci. Remote Sens.* **2016**, *54*, 7197–8009. [CrossRef]

42. Banda, F.; Dall, J.; Tebaldini, S. Single and Multipolarimetric P-Band SAR Tomography of Subsurface Ice Structure. *IEEE Trans. Geosci. Remote Sens.* **2016**, *54*, 2832–2845. [CrossRef]

43. Silva, W.F.; Rudorff, B.F.T.; Formaggio, A.R.; Paradella, W.R.; Mura, J.C. Discrimination of agricultural crops in a tropical semi-arid region of Brazil based on L-band polarimetric airborne SAR data. *ISPRS J. Photogramm. Remote Sens.* **2009**, *64*, 458–463. [CrossRef]

44. Lee, J.S.; Grunes, M.R.; Pottier, E.; Ferro-Famil, L. Unsupervised terrain classification preserving polarimetric scattering characteristics. *IEEE Trans. Geosci. Remote Sens.* **2004**, *42*, 722–731.

45. Alpers, W.; Holt, B.; Zeng, K. Oil spill detection by imaging radars: Challenges and pitfalls. *Remote Sens. Environ.* **2007**, *201*, 133–147. [CrossRef]

46. Rombach, M.; Fernandes, A.C.; Luebeck, D.; Moreira, J. Newest technology of mapping by using airborne interferometric synthetic aperture radar systems. In Proceedings of the IEEE International Geoscience and Remote Sensing Symposium, Toulouse, France, 21–25 July 2003.

47. Papa, C.; Alberti, G.; Salzillo, G.; Palmese, G.; Califano, D.; Ciofaniello, L.; Daniele, M.; Facchinetti, C.; Longo, F.; Formaro, R.; et al. Design and Validation of a Multimode Multifrequency VHF/UHF Airborne Radar. *IEEE Geosci. Remote Sens. Lett.* **2014**, *11*, 1260–1264. [CrossRef]

48. Rosa, R.A.S.; Fernandes, D.; Barreto, T.L.M.; Wimmer, C.; Nogueira, J.B. Deforestation detection in Amazon rainforest with multitemporal X-band and p-band sar images using cross-coherences and superpixels. In Proceedings of the IEEE International Geoscience and Remote Sensing Symposium, Fort Worth, TX, USA, 23–28 July 2017.

49. Available online: https://www.geomatics.metasensing.com/airborne-sar (accessed on 20 June 2019).

50. Wheeler, K.; Hensley, S. The GeoSAR airborne mapping system. In Proceedings of the Record of the IEEE 2000 International Radar Conference, Alexandria, VA, USA, 12 May 2000.

51. Walker, B.; Sander, G.; Thompson, M.; Burns, B.; Fellerhoff, R.; Dubbert, D. A High-Resolution, Four-Band SAR Testbed with Real-Time Image Formation. In Proceedings of the International Geoscience and Remote Sensing Symposium, Lincoln, NE, USA, 31–31 May 1996.

52. Dubois-Fernandez, P.; du Plessis, O.R.; le Coz, D.; Dupas, J.; Vaizan, B.; Dupuis, X.; Cantalloube, H.; Coulombeix, C.; Titin-Schnaider, C.; Dreuillet, P.; et al. The ONERA RAMSES SAR system. In Proceedings of the International Geoscience and Remote Sensing Symposium, Toronto, ON, Canada, 24–28 June 2002.

53. Baqué, R.; Bonin, G.; du Plessis, O.R. The airborne SAR-system: SETHI airborne microwave remote sensing imaging system. In Proceedings of the International Radar Conference "Surveillance for a Safer World", Bordeaux, France, 12–16 October 2009.

54. Imhoff, M.L.; Johnson, P.; Holford, W.; Hyer, J.; May, L.; Lawrence, W.; Harcombe, P. BioSAR^{TM}: An inexpensive airborne VHF multiband SAR system for vegetation biomass measurement. *IEEE Trans. Geosci. Remote Sens.* **2000**, *38*, 1458–1462. [CrossRef]

55. Ulander, L.M.H.; Blom, M.; Flood, B.; Follo, P.; Frolind, P.-O.; Gustavsson, A.; Jonsson, T.; Larsson, B.; Murdin, D.; Pettersson, M.; et al. Development of the ultra-wideband LORA SAR operating in the VHF/UHF-band. In Proceedings of the International Geoscience and Remote Sensing Symposium, Toulouse, France, 21–25 July 2003.
56. Zaugg, E.; Edwards, M.; Long, D.; Stringham, C. Developments in compact high-performance synthetic aperture radar systems for use on small Unmanned Aircraft. In Proceedings of the Aerospace Conference, Big Sky, MT, USA, 5–12 March 2011.
57. Dzenkevich, A.V.; Vostrov, E.A.; Mel'nikov, L.J.; Volkov, V.G.; Plyuschev, V.A.; Manakov, V.J. IMARC-multifrequency multipolarization airborne SAR system. In Proceedings of the Radar 97, Edinburgh, UK, 14–16 October 1997.
58. Pinheiro, M.; Reigber, A.; Scheiber, R.; Prats-Iraola, P.; Moreira, A. Generation of highly accurate DEMs over flat areas by means of dual-frequency and dual-baseline airborne SAR interferometry. *IEEE Trans. Geosci. Remote Sens.* **2018**, *56*, 4361–4390. [CrossRef]
59. Eisenburger, D.; Krellmann, Y.; Lentz, H.; Triltzsch, G. Stepped-frequency radar system in gating mode: An experiment as a new helicopter-borne GPR system for geological applications. In Proceedings of the IEEE International Geoscience and Remote Sensing Symposium, Boston, MA, USA, 7–11 July 2008.
60. Richards, M.A.; Scheer, J.A.; Holm, W.A. *Principles of Modern Radar: Basic Principles*; SciTech Publishing: Raleigh, NC, USA, 2010.
61. *IEEE Standard Definitions of Terms for Antennas*; IEEE Std 145-1983; The Institute of Electrical and Electronics Engineers: New York, NY, USA, 1983.
62. Moreira, A.; Prats-Iraola, P.; Younis, M.; Krieger, G.; Hajnsek, I.; Papathanassiou, K.P. A tutorial on synthetic aperture radar. *IEEE Geosci. Remote Sens. Mag.* **2013**, *1*, 6–43. [CrossRef]
63. Franceschetti, G.; Lanari, R. *Synthetic Aperture Radar Processing*; CRC PRESS: New York, NY, USA, 1999.
64. Costanzo, S.; Di Massa, G.; Costanzo, A.; Borgia, A.; Papa, C.; Alberti, G.; Salzillo, G.; Palmese, G.; Califano, D.; Ciofanello, L.; et al. Multimode/Multifrequency Low Frequency Airborne Radar Design. *J. Electr. Comput. Eng.* **2013**, *2013*, 12. [CrossRef]
65. Richards, M.A. *Fundamentals of Radar Signal Processing*; Tata McGraw-Hill Education: New York, NY, USA, 2005.
66. Gennarelli, G.; Ludeno, G.; Catapano, I.; Soldovieri, F.; Alberti, G.; Califano, D.; Ciofaniello, L.; Palmese, G.; Papa, C.; Pica, G.; et al. An improved airborne VHF radar sounder for ice and desert exploration. In Proceedings of the 17th International Conference on Ground Penetrating Radar (GPR), Rapperswil, Switzerland, 18–21 June 2018.
67. Chew, W.C. *Waves and Fields in Inhomogeneous Media*, 2nd ed.; IEEE: New York, NY, USA, 1995.
68. Bertero, M.; Boccacci, P. *Introduction to Inverse Problems in Imaging*; CRC PRESS: New York, NY, USA, 1998.
69. Persico, R.; Ludeno, G.; Soldovieri, F.; De Coster, A.; Lambot, S. Two-dimensional linear inversion of GPR data with a shifting zoom along observation line. *Remote Sens.* **2017**, *9*, 10. [CrossRef]
70. Gennarelli, G.; Catapano, I.; Soldovieri, F. Reconstruction capabilities of down-looking airborne GPRs: The single frequency case. *IEEE Trans. Comput. Imaging* **2017**, *3*, 917–927. [CrossRef]
71. Persico, R. *Introduction to Ground Penetrating Radar: Inverse Scattering and Data Processing*; John Wiley & Sons: Hoboken, NJ, USA, 2014.
72. Moreira, A.; Huang, Y. Airborne SAR Processing of Highly Squinted Data Using a Chirp Scaling Approach with Integrated Motion Compensation. *IEEE Trans. Geosci. Remote Sens.* **1994**, *32*, 1029–1040. [CrossRef]
73. Fornaro, G. Trajectory Deviations in Airborne SAR. Analysis and Compensation. *IEEE Trans. Aerosp. Electron. Syst.* **1999**, *35*, 997–1009. [CrossRef]
74. Cafforio, C.; Prati, C.; Rocca, F. SAR data focusing using seismic migration techniques. *IEEE Trans. Aerosp. Electron. Syst.* **1991**, *27*, 194–207. [CrossRef]
75. Fornaro, G.; Franceschetti, G.; Perna, S. On Center-Beam Approximation in SAR Motion Compensation. *IEEE Geosci. Remote Sens. Lett.* **2006**, *3*, 276–280. [CrossRef]
76. Fornaro, G.; Franceschetti, G.; Perna, S. Motion Compensation Errors: Effects on the Accuracy of Airborne SAR Images. *IEEE Trans. Aerosp. Electron. Syst.* **2005**, *41*, 1338–1352. [CrossRef]
77. Soumekh, M. *Synthetic Aperture Radar Signal Processing with MATLAB Algorithms*; Wiley: Hoboken, NJ, USA, 1999.
78. Lee, J.S.; Pottier, E. *Polarimetric Radar Imaging: From Basics to Applications*; CRC Press: New York, NY, USA, 2009.
79. Van Zyl, J.J. Unsupervised classification of scattering behavior using radar polarimetry data. *IEEE Trans. Geosci. Remote Sens.* **1989**, *27*, 36–45. [CrossRef]

80. Iodice, A.; Natale, A.; Riccio, D. Polarimetric Two-Scale Model for Soil Moisture Retrieval via Dual-Pol HH-VV SAR Data. *IEEE J. Sel. Top. Appl. Earth Obs. Remote Sens.* **2013**, *6*, 1163–1171. [CrossRef]
81. D'Alessandro, M.M.; Tebaldini, S. Digital Terrain Model Retrieval in Tropical Forests Through P-Band SAR Tomography. *IEEE Trans. Geosci. Remote Sens.* **2019**, 1–8. [CrossRef]

Article

Focusing High-Resolution Highly-Squinted Airborne SAR Data with Maneuvers

Shiyang Tang [1,*], Linrang Zhang [1] and Hing Cheung So [2]

[1]   National Lab. of Radar Signal Processing, Xidian University, Xi'an 710071, China; lrzhang@xidian.edu.cn
[2]   Department of Electronic Engineering, City University of Hong Kong, Hong Kong, China;
       hcso@ee.cityu.edu.hk
*   Correspondence: sytang@xidian.edu.cn; Tel.: +86-153-1992-2377

Received: 24 April 2018; Accepted: 29 May 2018; Published: 1 June 2018

Abstract: Maneuvers provide flexibility for high-resolution highly-squinted (HRHS) airborne synthetic aperture radar (SAR) imaging and also mean complex signal properties in the echoes. In this paper, considering the curved path described by the fifth-order motion parameter model, effects of the third- and higher-order motion parameters on imaging are analyzed. The results indicate that the spatial variations distributed in range, azimuth, and height directions, have great impacts on imaging qualities, and they should be eliminated when designing the focusing approach. In order to deal with this problem, the spatial variations are decomposed into three main parts: range, azimuth, and cross-coupling terms. The cross-coupling variations are corrected by polynomial phase filter, whereas the range and azimuth terms are removed via Stolt mapping. Different from the traditional focusing methods, the cross-coupling variations can be removed greatly by the proposed approach. Implementation considerations are also included. Simulation results prove the effectiveness of the proposed approach.

Keywords: high-resolution; highly-squinted; maneuvers; fifth-order motion parameter model; spatial variation

---

1. Introduction

In recent years, there have been tremendous studies on synthetic aperture radar (SAR). As an active sensor, SAR is able to work day and night under all weather conditions [1–3]. In addition, SAR can operate at different frequencies and view angles in different polarimetric modes. This feature makes the SAR a flexible and effective tool for information retrieval [4–6]. With the advancement of SAR, high resolution and highly squint angle have potential to provide more information about the surface structure. Moreover, the SAR platform is capable of flying along a curved path to realize different applications, to the extent that the model assumption of the rectilinear path no longer holds. This scenario occurs in SAR systems on aircraft platforms because of various factors such as rugged topography, atmospheric turbulence, and intended maneuvers [7–10]. Characteristics of a curved path differ from those of a uniform linear motion. Major peculiarities exist in its motion parameters, non-uniform spatial-intervals, and three-dimensional (3-D) spatial geometric model. Thus, the traditional imaging algorithms based on straight trajectory and hyperbolic range model (HRM) may be invalid. In order to guarantee the imaging qualities, both the range model and the imaging algorithm may need to change. In addition, the maneuvers will greatly affect the spatial variations in both the range and azimuth directions, particularly for the high-resolution highly-squinted (HRHS) SAR.

In the literature, the fourth-order Doppler range model (FORM) [11–13], modified equivalent squint range model (MESRM) [14], advanced hyperbolic range equation (AHRE) [15], equivalent range model (ERM) [16], and modified ERM (MERM) [17] for spaceborne or airborne SAR have been proposed to describe the curved path. Compared with the conventional HRM, these models introduce acceleration into the range model, which makes the descriptions of characteristics, including Doppler bandwidth, cross-coupling phase, and two-dimensional (2-D) spatial variations, of the raw-data more accurate. However, they only consider the acceleration term and ignore the higher-order motion parameters, which limit their applications for high-accuracy imaging. In reality, the maneuvers cannot always be controlled only by constant velocity and acceleration, thus the higher-order motion parameters are needed [18]. If this problem cannot be well solved, it may strongly impair the final image quality in terms of geometric distortion and radiometric resolution losses for HRHS SAR [5]. Thus, profound research on the geometrical model is still necessary.

Concerning the focusing algorithm for the SAR with maneuvers, methods performed in the frequency domain include the range-Doppler algorithm (RDA) [19], chirp scaling algorithm (CSA) [20], omega-K algorithm (OKA) [21], and their extensions [11–13,22–25]. Eldhuset [11,12] suggests a fourth-order processing algorithm by 2-D exact transfer function (ETF) for spaceborne SAR with curved orbit. However, this work ignores the 2-D spatial variation of the azimuth modulation phase and results in defocusing in the azimuth edge regions. Luo et al. [13] and Wang et al. [14] respectively propose a modified RDA and a modified CSA, which can greatly remove the cross-coupling terms brought by the curved path. However, the spatial variations of the acceleration have not been considered. Li et al. [23] propose a frequency-domain algorithm (FDA) for the small-aperture highly-squinted airborne SAR with maneuvers. With expanding the azimuth time in a small aperture, the azimuth spatial variation of the stripmap SAR can be eliminated. However, the residual errors increase greatly with the aperture (resolution). Moreover, the neglected range and vertical spatial variations of the azimuth modulation phase cannot be ignored for the HRHS SAR with maneuvers. The wavenumber domain algorithm [17] and OKA [22] for the HRHS SAR with curved path are proposed based on different modified equivalent range models, which can avoid using the method of series reversion (MSR) to achieve the 2-D spectrum. However, the residual spatial variations caused by approximations would lead to deteriorations in the final image. The 2-D keystone transform algorithms (KTAs) are developed in [26] based on the 2-D Taylor series expansion and they can greatly remove the spatial variations of the high-resolution spaceborne SAR. The errors introduced by the 2-D Taylor expansion can be ignored for the spaceborne SAR but not for the HRHS SAR. Generally, the 2-D spatial variations are not eliminated entirely by [11–24] performed in frequency domain and have great impacts on the final image result. Wu et al. [27] propose a hybrid correlation algorithm (HCA) for the curved flight path, which treats the 2-D correlation by a combination of frequency-domain fast correlation in azimuth dimension and time-domain convolution in the range dimension. Furthermore, back projection algorithm (BPA) and fast factorized BPA (FFBPA) [28–31] have been suggested. However, in terms of computational burden, the HCA and BPA are not always the best choices compared with the frequency-domain algorithms. The polar format algorithm (PFA) [5,32] can be used for the three-dimensional (3-D) acceleration cases. However, the depth of field is seriously affected by the wavefront curvature, and it must be extended by subaperture technique for the quadratic phase error (QPE) compensation. Thus, further studies are still required for the HRHS SAR with maneuvers.

In this paper, the fifth-order motion parameter model is introduced and the problems are discussed for the HRHS SAR with maneuvers, which are the important factors that demand attention in imaging design. Our analyses suggest that the spatial variations in arbitrary direction brought by the third- and higher-order motion parameters cannot be ignored. Employing the Taylor series expansions with multi-variables, we decompose the spatial variations into three parts—i.e., range, azimuth, and cross-coupling terms—with a high accuracy. Then, according to the properties of decomposed phases, the polynomial phase filter and Stolt mapping with interpolations are performed to remove the cross-coupling and range/azimuth spatial variations, respectively. Unlike the traditional focusing algorithms [11–15,23,24], the cross-coupling spatial variations, which are always ignored in low-resolution case, are corrected for the HRHS SAR with maneuvers. Implementation considerations, including simplified processing and constraint on scene extent are also studied.

The rest of this paper is organized as follows. The signal model of HRHS SAR with maneuvers is investigated and the confronting problems are presented in Section 2. In Section 3, our imaging approach is presented. Implementation considerations are provided in Section 4. Numerical simulation results are given to validate the proposed approach in Section 5. Conclusions are drawn in Section 6.

2. Modeling and Motivation

2.1. Modeling

Figure 1 shows the geometric model of the HRHS SAR with maneuvers. The projection of radar location on the ground is assumed to be the origin of Cartesian coordinates O-XYZ. Assuming that points C and A are respectively the central reference point (CRP) and an arbitrary point on the scene, *l* is the flight path of the platform, point Q is the position of the platform at the aperture center moment (ACM), and r_c and r_A are respectively the position vectors of platform to points C and A at ACM.

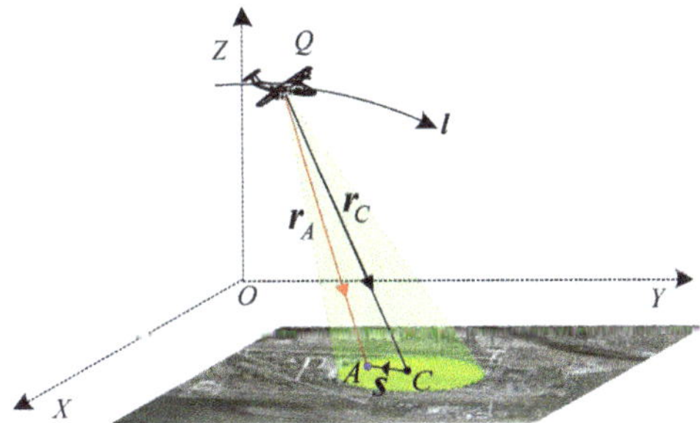

Figure 1. HRHS SAR imaging geometry with maneuvers.

According to the imaging geometry shown in Figure 1 and the kinematics equation of the platform, the instantaneous slant range history $|r(\eta)|$ corresponding to arbitrary point A can be expressed as

$$|r(\eta)| = \left| r_A - v\eta - a\eta^2/2 - b\eta^3/6 - c\eta^2/24 - d\eta^2/120 \right| \tag{1}$$

where $|\cdot|$ is the symbol of absolute value, v is the velocity vector, and a is the acceleration vector, while b, c, and d are the third-, fourth-, and fifth- order motion parameter vectors in the motion Equation (1), respectively. It is obvious that the range history $|r(\eta)|$ is an equation with higher-order terms shown as a flat-top shape. It is difficult to derive the 2-D spectrum directly using Equation (1) based on the principle of stationary point (POSP); therefore, the traditional SAR processing methods cannot be applied directly. One way to treat the complex fifth-order motion parameter model (FMPM) is to expand it into a power series in azimuth time as

$$|r(\eta)| = \sum_0 \frac{1}{n!}\mu_n\eta^n \tag{2}$$

where the first six coefficients are

$$\mu_0 = \sqrt{\langle r_A, r_A \rangle} \tag{3}$$

$$\mu_1 = \mu_0^{-1} \cdot [-\langle r_A, v \rangle] \tag{4}$$

$$\mu_2 = \mu_0^{-1} \cdot \left[(-\langle r_A, a \rangle + \langle v, v \rangle) - \mu_1^2 \right] \tag{5}$$

$$\mu_3 = \mu_0^{-1} \cdot [(-\langle r_A, b \rangle + 3\langle v, a \rangle) - 3\mu_1\mu_2] \tag{6}$$

$$\mu_4 = \mu_0^{-1} \cdot \left[\left(-\langle \boldsymbol{r}_A, \boldsymbol{c} \rangle + 4\langle \boldsymbol{v}, \boldsymbol{b} \rangle + 3\langle \boldsymbol{a}, \boldsymbol{a} \rangle \right) - 3\mu_2^2 - 4\mu_1\mu_3 \right] \tag{7}$$

$$\mu_5 = \mu_0^{-1} \cdot \left[\left(-\langle \boldsymbol{r}_A, \boldsymbol{d} \rangle + 5\langle \boldsymbol{v}, \boldsymbol{c} \rangle + 10\langle \boldsymbol{a}, \boldsymbol{b} \rangle \right) - 10\mu_2\mu_3 - 5\mu_1\mu_4 \right] \tag{8}$$

In Equation (1), μ_0, μ_1, and μ_2 are respectively the slant range, Doppler centroid, and Doppler frequency modulation (FM) of the point A. μ_3, μ_4, and μ_5 are the higher-order terms which have great impact on the final image qualities and cannot be ignored.

By using the range history expressed in Equation (1), we obtain the received echo as

$$S_0(t_r, \eta) = \varepsilon_0 w_r(t_r) w_a(\eta) \exp(-j4\pi f_c |r(\eta)|/c) \exp\left(j\pi\gamma[t_r - 2|r(\eta)|/c]^2 \right) \tag{9}$$

where t_r is the range fast time, c is the speed of light, f_c and γ are the carrier frequency and FM rate of the transmitted signal respectively, ε_0 is the complex scattering coefficient, and $w_r(\cdot)$ and $w_a(\cdot)$ are the range and azimuth windows in time domain.

Based on Equation (2), Equation (9) is transformed into the range frequency domain using the POSP after range compression, i.e.,

$$S_1(f_r, \eta) = \varepsilon_0 w_r(f_r) w_a(\eta) \exp\left(-j\frac{4\pi(f_c + f_r)}{c} |r(\eta)| \right) \tag{10}$$

where f_r is the range frequency and $\omega_r(\cdot)$ is the range window in frequency domain.

2.2. Motivation

(1) Error Analysis of FMPM: Traditionally, $\boldsymbol{v}$ and $\boldsymbol{a}$ are always taken into consideration for the SAR with maneuvers. However, in the case of HRHS SAR, this second-order motion parameter equation is insufficient and it will greatly deteriorate the final image results and limit the scene size. In this work, the higher-order motion parameter vectors, namely, $\boldsymbol{b}$, $\boldsymbol{c}$, and $\boldsymbol{d}$ are exploited to improve the accuracy of the flight path description.

Employing the parameters listed in Table 1, Figure 2a,b respectively show the phase errors and spatially variant errors caused by the motion parameter vectors $\boldsymbol{b}$, $\boldsymbol{c}$, and $\boldsymbol{d}$. The unit of the contour maps is π. Note that spatially variant errors brought by $\boldsymbol{d}$ can be ignored, whereas both the phase errors by $\boldsymbol{b}$, $\boldsymbol{c}$, and $\boldsymbol{d}$ and the spatially variant errors by $\boldsymbol{b}$ and $\boldsymbol{c}$ cannot. Figure 3a,b respectively show the spatially variant errors with different azimuth resolutions in range and azimuth directions. Clearly, the effects of $\boldsymbol{b}$ and $\boldsymbol{c}$ must be taken into consideration for the high-resolution cases whereas that of $\boldsymbol{d}$ is negligible. Figure 4a,b respectively show the spatially variant errors with different squint angles in range and azimuth directions. The maximum errors at large squint angles are far larger than $\pi/4$ introduced by $\boldsymbol{b}$ and $\boldsymbol{c}$. The impacts brought by $\boldsymbol{d}$ are still small enough and can be ignored. According to the above analyses, the spatial variations brought by the motion parameter vectors $\boldsymbol{b}$ and $\boldsymbol{c}$ should be considered.

Table 1. Parameter settings.

Motion Parameter	Value	System Parameter	Value
Radar Position at ACM	$(0, 0, 10)$ km	Carrier Frequency	17 GHz
Reference Position	$(12.68, 26, 0)$ km	Pulse Bandwidth	500 MHz
Velocity v	$(0, 170, -10)$ m/s	Sampling Frequency	620 MHz
Acceleration a	$(1.2, 1.73, -1.4)$ m/s^2	Squint Angle	60°
Third-order Parameter b	$(-0.09, 0.11, -0.14)$ m/s^3	Azimuth Resolution	0.242 m
Fourth-order Parameter c	$(0.005, 0.007, 0.003)$ m/s^4	Scene Size (Range × Azimuth)	1.6 × 1.6 km

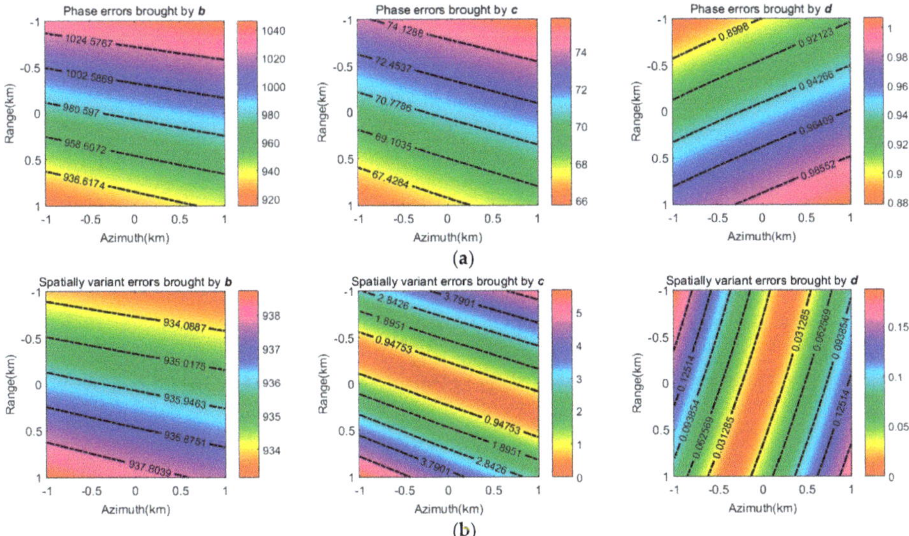

Figure 2. Impacts of motion parameter vectors b, c, and d on imaging results. (**a**) Phase errors and (**b**) spatially variant errors.

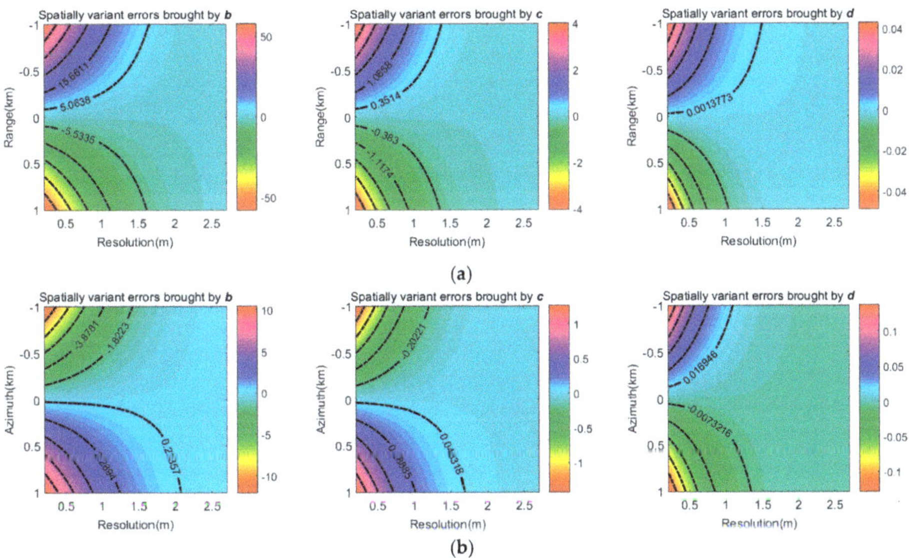

Figure 3. Spatial variations brought by motion parameter vectors, b, c, and d. (**a**) Range spatial variations with respect to range distance and resolution. (**b**) Azimuth spatially variant errors with respect to range distance and resolution.

(2) Irregularly Spatial Variation Distributions: As analyzed in the above part, the spatial variation should be taken into consideration for the HRHS SAR with maneuvers to achieve a high quality image and it exists in all the targets, with different range curvatures, with respect to the reference one on the scene. To better understand the existing spatial variations of the targets on the scene, an illustration is

provided in Figure 5, where T0 is the reference point, T1 and T2 are the targets that respectively have the same azimuth and range cells as those of T0, and T3 is the target that has different position as that of T0. Basically, there are three kinds of spatial variations irregularly distributed in the ground scene: range, azimuth, and 2-D cross-coupling spatial variations, as shown in Figure 5a. The range or azimuth spatial variations are traditionally processed whereas the 2-D cross-coupling one, which is irregularly distributed on the ground scene as shown in Figure 2b, is ignored. Moreover, the spatial variations exist in both azimuth time and azimuth frequency domains, as shown in Figure 5b. The curved non-parallel time-frequency diagrams (TFDs) indicate that spatial variations of range cell migration (RCM) and secondary range cell (SRC) in either azimuth time or azimuth frequency domains should be compensated when designing the focusing algorithm.

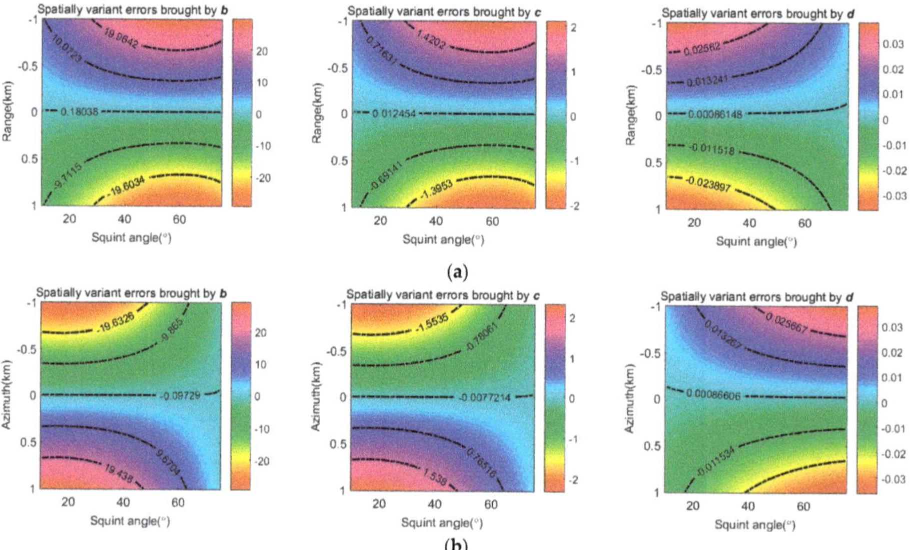

Figure 4. Spatial variations brought by motion parameter vectors *b* , *c*, and *d*. (**a**) Range spatial variations with respect to range distance and squint angle. (**b**) Azimuth spatially variant errors with respect to range distance and squint angle.

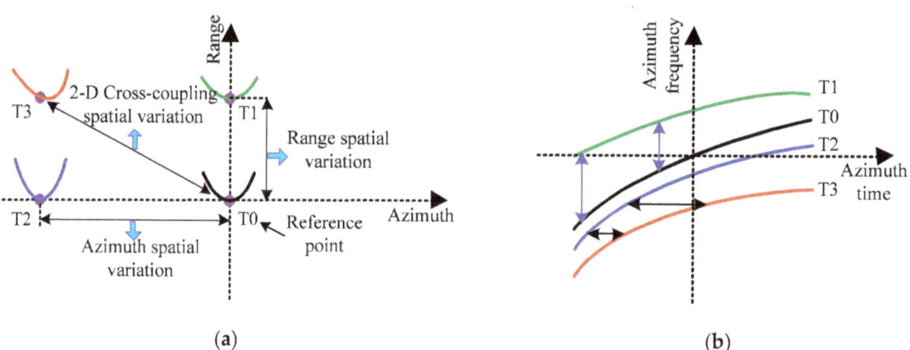

Figure 5. Spatial variations distribution. (**a**) Spatial variations distributed in ground scene. (**b**) TFDs of targets T0, T1, T2, and T3.

3. Imaging Approach

We propose an imaging approach that combines polynomial phase filter and Stolt mapping. The first step is to eliminate the cross-coupling spatial variations in the second-, third-, and fourth-order phases via polynomial phase filtering. The second step is to correct the spatial variations in range and azimuth directions through the Stolt mapping.

3.1. 2-D Cross-Coupling Spatial Variation Elimination

We first construct a polynomial phase filter to eliminate the second-, third-, and fourth-order cross-coupling spatially variant terms corresponding to the Doppler centroid μ_1, which is a key function for the whole imaging approach

$$H_1(\eta) = \exp\left\{ j\frac{4\pi(f_r + f_c)}{c}\left|r_{ref}(\eta)\right| \right\} \exp\left\{ -j\frac{4\pi(f_r + f_c)}{c}\sum_{n=2}^{4}\frac{1}{n!}\chi_n\eta^n \right\} \tag{11}$$

where the coefficients χ_2, χ_3, and χ_4 are to be determined. $\left|r_{ref}(\eta)\right|$ is the range history of the reference point and the subscript '*ref*' denotes the reference target. The first phase term is used for the bulk compensation and it can greatly decrease the impacts brought by the high squint angle and spatially invariant terms. The second term, i.e., polynomial phase filter, aims to eliminate the cross coupling spatial variation.

Multiplying Equation (11) by Equation (10) and transforming the result into 2-D frequency domain by using POSP, we can then obtain (see Appendix A)

$$S_2(f_r, f_\eta) \approx \varepsilon_0 w_r(f_r)w_a(f_\eta)\exp\left\{ j\pi\frac{4(f_r + f_c)}{c}\left(k_0 + k_1 f_k + k_2 f_k^2 + k_3 f_k^3 + k_4 f_k^4\right) \right\} \tag{12}$$

where $f_k = f_\eta c/[2(f_r + f_c)]$, with f_η being the azimuth frequency, $w_a(\cdot)$ is the azimuth window in frequency domain, while k_0 and k_1 are respectively the range and azimuth position terms and they have no impact on the imaging results. Thus, we decompose the higher-order spatially variant phase terms corresponding to k_2, k_3, and k_4 into four parts: range, azimuth, and cross-coupling spatially variant terms, as well as spatially invariant term, i.e.,

$$S_3(f_r, f_\eta) \approx \varepsilon_0 w_r(f_r)w_a(f_\eta)\exp\left\{ j\left[\varphi_{ran}(f_r, f_\eta) + \varphi_{azi}(f_r, f_\eta) + \varphi_{cou}(f_r, f_\eta) + \varphi_{con}(f_r, f_\eta)\right] \right\} \tag{13}$$

where $\varphi_{ran}(f_r, f_\eta)$, $\varphi_{azi}(f_r, f_\eta)$, and $\varphi_{cou}(f_r, f_\eta)$ are respectively the range, azimuth, and cross-coupling spatially variant terms, and $\varphi_{con}(f_r, f_\eta)$ is the spatially invariant term (see Appendix B)

$$\varphi_{ran}(f_r, f_\eta) = \frac{4\pi(f_r + f_c)}{c}\left(1 + \sum_{i=2}^{4} p_i(\chi_2, \chi_3, \chi_4)f_k^i\right) \cdot k_0 \tag{14}$$

$$\varphi_{azi}(f_r, f_\eta) = \frac{4\pi(f_r + f_c)}{c}\left(f_k + \sum_{i=2}^{4} q_i(\chi_2, \chi_3, \chi_4)f_k^i\right) \cdot k_1 \tag{15}$$

$$
\begin{aligned}
\varphi_{cou}(f_r, f_\eta) = \frac{4\pi(f_r+f_c)}{c}\Big\{ & l_2(\chi_2, \chi_3, \chi_4)\left(\mu_1 - \mu_1^{ref}\right) \cdot f_k^2 \\
& +l_3(\chi_2, \chi_3, \chi_4)\left(\mu_1 - \mu_1^{ref}\right)^2 \cdot f_k^2 \\
& +l_4(\chi_2, \chi_3, \chi_4)\left(\mu_1 - \mu_1^{ref}\right) \cdot f_k^3 \Big\}
\end{aligned}
\tag{16}
$$

$$\varphi_{con}(f_r, f_\eta) = \frac{4\pi(f_r + f_c)}{c}\sum_{i=2}^{4} z_i(\chi_2, \chi_3, \chi_4)f_k^i \tag{17}$$

where all $p_i(\chi_2, \chi_3, \chi_4)$, $q_i(\chi_2, \chi_3, \chi_4)$, $l_i(\chi_2, \chi_3, \chi_4)$, and $z_i(\chi_2, \chi_3, \chi_4)$ $(i = 2, 3, 4)$ are spatially invariant coefficients which are derived by phase decompositions with the use of the gradient method [33] in the 3-D geographical space. The Taylor series expansion with multi-variables employed in Appendix B has higher accuracy than that with one variable [33,34], which avoids deterioration in the final imaging result.

To remove the cross-coupling spatial variations brought by μ_1, the coefficients $l_i(\chi_2, \chi_3, \chi_4)$ $(i = 2, 3, 4)$ in Equation (16) should be set to zero. Thus, one can get the following three equations with three unknowns, i.e.,

$$\begin{cases} l_2(\chi_2, \chi_3, \chi_4) = 0 \\ l_3(\chi_2, \chi_3, \chi_4) = 0 \\ l_4(\chi_2, \chi_3, \chi_4) = 0 \end{cases} \tag{18}$$

The solving process of Equation (18) for χ_2, χ_3, and χ_4 is also provided in Appendix B. Substituting the solutions of χ_2, χ_3, and χ_4 into $p_i(\chi_2, \chi_3, \chi_4)$, $q_i(\chi_2, \chi_3, \chi_4)$, and $z_i(\chi_2, \chi_3, \chi_4)$ $(i = 2, 3, 4)$ respectively in Equations (14), (15), and (17), accurate expressions of $\varphi_{ran}(f_r, f_\eta)$, $\varphi_{azi}(f_r, f_\eta)$, and $\varphi_{con}(f_r, f_\eta)$ are obtained.

Illustration of the processing scheme is provided in Figure 6 to better understand the whole procedure. The TFDs of the cross-coupling spatial variations are shown in Figure 6a. After bulk compensation, the cross-coupling spatial variations are greatly weakened as shown in Figure 6b. Then, the TFD of the polynomial phase filter, presented in Figure 6c, is applied. The cross-coupling spatial variations are corrected in the 2-D frequency domain and the result is shown in Figure 6d. It is worth noting that the polynomial phase filter is more like a perturbation function in the traditional azimuth nonlinear CS (ANLCS) with similar solving process [23,24,35,36]. The difference is that the perturbation function in ANLCS only eliminates the azimuth spatially variant phase brought by range walk correction in stripmap mode whereas the polynomial phase filter can greatly remove the cross-coupling one which is ignored in traditional ANLCS. Moreover, the polynomial phase filter can also process the height spatial variations due to the topography variations.

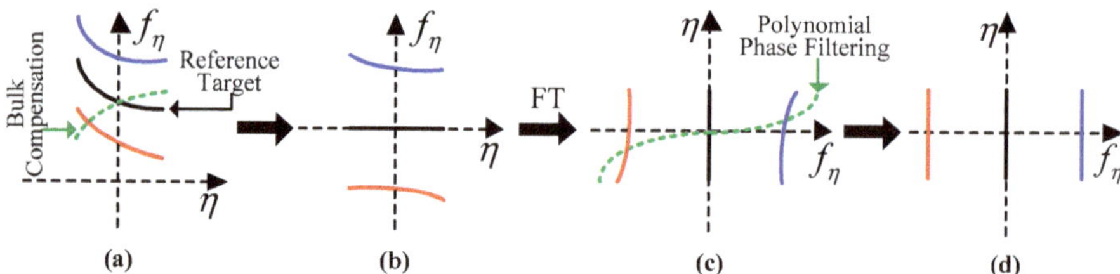

Figure 6. Illustration of cross-coupling spatial variation elimination by TFDs of three targets with different positions. (**a**,**b**) TFDs before and after bulk compensation, respectively; (**c**,**d**) TFDs before and after polynomial phase filtering.

3.2. Range and Azimuth Spatial Variation Elimination

After polynomial phase filtering, the possible azimuth spectrum aliasing should be taken into consideration. Noting two main aspects: (1) the azimuth bandwidth is greatly affected by the motion parameters; (2) the Doppler FM becomes $\mu_2 + \chi_2$ after polynomial phase filtering, which means that the azimuth bandwidth may have a big change, aliasing should be eliminated before the azimuth Fourier transform (FT) of the signal. The corresponding solution has been discussed in [16] and one can use it for efficient data preprocessing. It is worth mentioning that the expressions of the 2-D spectrum before and after the preprocessing are similar except for azimuth frequency variable f_η in Equation (13).

After the cross-coupling spatial variation elimination, the echo signal is expressed as

$$S_4\left(f_r, f_\eta\right) \approx \varepsilon_0 w_r(f_r) w_a\left(f_\eta\right) \exp\left\{ j\left[\varphi_{ran}\left(f_r, f_\eta\right) + \varphi_{azi}\left(f_r, f_\eta\right) + \varphi_{con}\left(f_r, f_\eta\right)\right]\right\} \tag{19}$$

In Equation (19), the spatially invariant phase term can be compensated by

$$H_2\left(f_r, f_\eta\right) = \exp\left(-j\varphi_{con}\left(f_r, f_\eta\right)\right) \tag{20}$$

The first- and second-terms in Equation (19) are respectively the range and azimuth modulation phases, which determine the range and azimuth positions. In this case, the ideal solution is to perform separable interpolation respectively in the range and azimuth directions to remove the corresponding spatial variations. The formulas of the interpolation are expressed as

$$\frac{4\pi(f_r + f_c)}{c}\left(1 + \sum_{i=2}^{4} p_i(\chi_2, \chi_3, \chi_4)f_k^i\right) \rightarrow 4\pi\left(f_r' + f_c\right)/c \tag{21}$$

$$\frac{4\pi(f_r + f_c)}{c}\left(f_k + \sum_{i=2}^{4} q_i(\chi_2, \chi_3, \chi_4)f_k^i\right) \rightarrow 2\pi f_\eta' \tag{22}$$

where f_r' and f_η' are, respectively, the new range and azimuth frequencies after interpolations. These substitutions are viewed as a Stolt mapping of $\left(f_r, f_\eta\right)$ into $\left(f_r', f_\eta'\right)$; thus, the echo signal becomes

$$S_5\left(f_r', f_\eta'\right) = \varepsilon_0 w_r(f_r') w_a\left(f_\eta'\right) \exp\left\{ j\left[4\pi\left(f_r' + f_c\right)/c \cdot k_0 + 2\pi f_\eta' \cdot k_1\right]\right\} \tag{23}$$

Clearly, a 2-D inverse FT (IFT) can be applied with Equation (23) to obtain a focused result, i.e.,

$$S_6(t_r, \eta) = \varepsilon_0 G_r G_a \mathrm{sinc}\{B_r[t_r - 2k_0/c]\} \cdot \mathrm{sinc}\{\Delta f_a[\eta - k_1]\} \tag{24}$$

where G_r and G_a denote, respectively, the range and azimuth compression gains, B_r is the bandwidth of the transmitted signal, and Δf_a is the azimuth bandwidth.

To illustrate the proposed algorithm, we consider a simple highly squinted flight geometry shown in Figure 1, with point target A. Figure 7 shows the results of the proposed algorithm by 2-D spectra of a target and the impulse response after compression. The solid lines in the first two rows represent phase contours. Figure 7a–c respectively show the support areas of 2-D cross-coupling, range, and azimuth spatially variant spectra after phase decomposition. RCM and SRC are generally very small, compared to the range bandwidth, but are exaggerated here to illustrate the effect of a target away from the reference point. The slightly curved phase contours indicate that the target is not properly focused. In Figure 7d, the phase is completely independent of range and azimuth frequencies after the polynomial phase filtering, which means that the 2-D cross-coupling spatially variant terms are eliminated entirely. The Stolt mappings of Figure 7b,c produce noticeable changes in phase contours, of which the lines are equally spaced and parallel as shown in Figure 7e,f. The echo data are well focused in range and azimuth directions respectively shown in Figure 7g,h after corresponding spatial variations being eliminated. Figure 7i shows the 2-D contour result of target.

3.3. Flowchart of Imaging Algorithm

The flowchart of the imaging approach is shown in Figure 8. By using the proposed approach, spatial variations including the range, azimuth, and height spatially variant phases are greatly removed for the HRHS SAR with maneuvers. It should be noted that the range history in (1) is not a general model. When the maximum phase error between the polynomials in (1) and the real supporting points is less than $\pi/4$, it has no impact on the imaging results. On the other hand, if the maximum phase error is larger than $\pi/4$, the final image would be deteriorated. One solution is to take higher-order motion parameters into consideration in the proposed approach to decrease the errors. The residual phase errors can also be compensated by using the autofocus techniques which have been clearly discussed in [37,38].

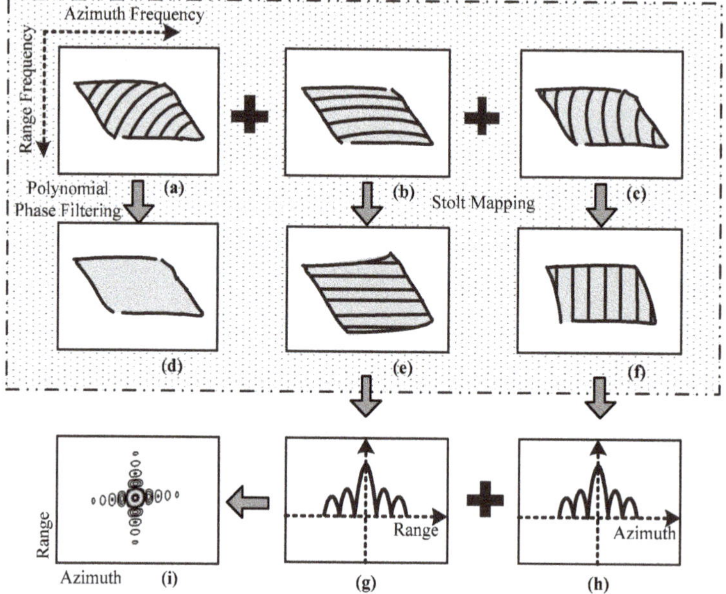

Figure 7. Illustration of proposed algorithm by 2-D spectra of a target and impulse response after compression. (**a**–**c**) in top row are 2-D spectra after phase decomposition, while (**d**–**f**) in the second row illustrate 2-D spectra after corresponding processing, and (**g**–**i**) in the bottom row are imaging results. The solid lines in the first two rows represent phase contours.

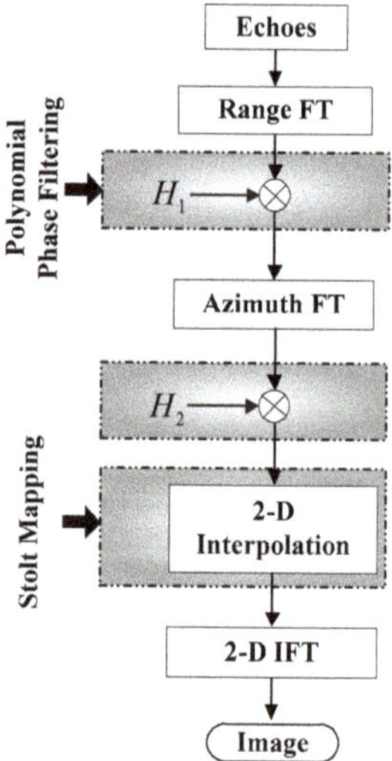

Figure 8. Flowchart of the proposed algorithm.

4. Implementation and Discussion

4.1. Simplified Processing

According to the mapping functions in Section 3.2, if the terms $4\pi(f_r + f_c)\left(1 + \sum_{i=2}^{4} p_i(\chi_2, \chi_3, \chi_4)f_k^i\right)/c$ in Equation (21) and $4\pi(f_r + f_c)\left(f_k + \sum_{i=2}^{4} q_i(\chi_2, \chi_3, \chi_4)f_k^i\right)/c$ in Equation (22) are sufficiently small, we can omit the corresponding interpolation to decrease the computational load of the whole imaging algorithm. In this subsection, a simplified processing method is suggested. In order to avoid the interpolation operation and to retain the image quality, the following two conditions must be satisfied

$$\max\left\{\left|4\pi(f_r + f_c)\left(1 + \sum_{i=2}^{4} p_i(\chi_2, \chi_3, \chi_4)f_k^i\right)/c\right|\right\} < \pi/4 \tag{25}$$

$$\max\left\{\left|4\pi(f_r + f_c)\left(f_k + \sum_{i=2}^{4} q_i(\chi_2, \chi_3, \chi_4)f_k^i\right)/c\right|\right\} < \pi/4 \tag{26}$$

and thus the impacts on the final image could be ignored. With simulation parameters in Table 1, the maximum phase errors of Equations (25) and (26) are 0.02 π and 0.7 π, respectively. Clearly, the azimuth interpolation is still necessary whereas the range one is not in this case. However, the results may not be generalizable. Thus, a judgment is added in the processes to determine whether the interpolation is necessary or not according to Equations (25) and (26). The judgment flowchart is given in Figure 9.

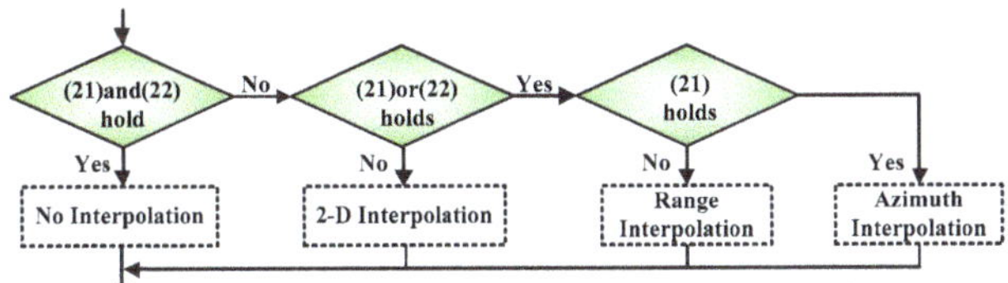

Figure 9. Judgment flowchart of the interpolation.

4.2. Constraint on Scene Extent

The scene extent is analyzed in this subsection. The scene size is mainly determined by the accuracy of the proposed approach. In the focusing step, approximations only occur in the phase decomposition. According to Equation (13), the phase error is derived as

$$\Phi_{res} = \frac{4\pi(f_r + f_c)}{c}\left(\sum_{i=2}^{4} k_i f_k^i - k_0 \sum_{i=2}^{4} p_i f_k^i - k_1 \sum_{i=2}^{4} q_i f_k^i\right) \tag{27}$$

To ensure the image quality, Φ_{res} should be less than $\pi/4$. By computing the Taylor series expansion $k_i(i = 0, 1, \cdots, 4)$ with respect to s and ignore the higher-order terms, i.e., $k_i = k_i^{ref} + \left\langle \nabla k_i^{ref}, s \right\rangle$, Φ_{res} can be rewritten as

$$\Phi_{res} \approx \frac{4\pi(f_r + f_c)}{c}\left(\sum_{i=2}^{4} \left\langle \nabla k_i^{ref}, s \right\rangle f_k^i - \left\langle \nabla k_0^{ref}, s \right\rangle \sum_{i=2}^{4} p_i f_k^i - \left\langle \nabla k_1^{ref}, s \right\rangle \sum_{i=2}^{4} q_i f_k^i\right) \tag{28}$$

where $k_i^{ref} = 0$. As $\Phi_{res} < \pi/4$, we obtain the scene sizes, i.e., $2s$, in both the horizontal and vertical directions. According to Equation (28), the decomposed errors increase with the focus depths in range, azimuth, and height directions. These phase errors could have negative effects on the SAR image formation when they are larger than $\pi/4$.

Utilizing airborne SAR simulation parameters in Table 1, Figure 10 shows the phase errors introduced by phase decompositions in range/azimuth, range/height, and azimuth/height planes. Clearly, the maximum phase errors in Figure 10a–c are all less than $\pi/4$, which means that the residual spatial variations in range, azimuth, and height directions after focusing are small enough and thus have negligible impact on the imaging qualities.

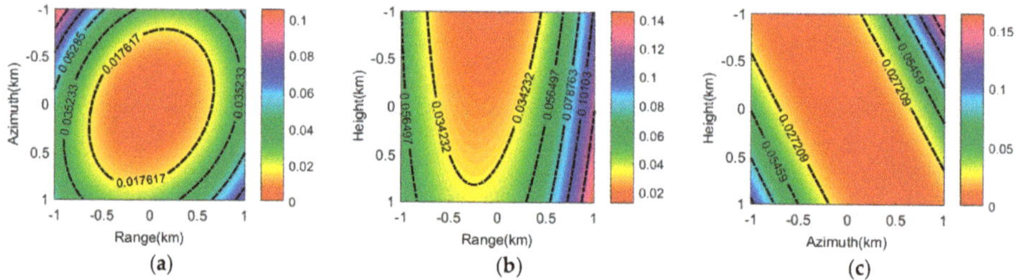

Figure 10. Phase errors in (**a**) range and azimuth, (**b**) range and height, and (**c**) azimuth and height planes.

5. Simulation Results

To prove the effectiveness of the proposed approach, simulation results are presented in this section.

5.1. Experiment 1

In this subsection, a spotlight mode SAR is simulated with a 3×3 dot-matrix being arranged in the simulation scene. The geometry of the scene is presented in Figure 11. The parameters are listed in Table 1.

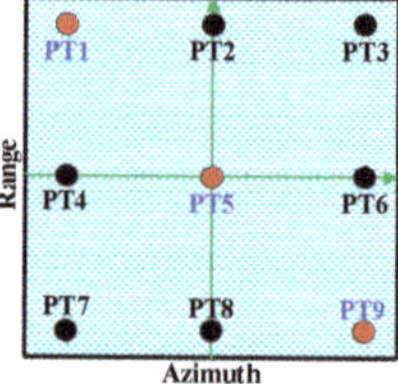

Figure 11. Ground scene for simulation.

Case 1: The motion parameters a, b, and c in this case are listed in Table 1. Simulation results without considering b and c are respectively used for comparisons. Moreover, the results by the FDA [16] are included. Figure 12 shows the comparative results of targets PT1, PT5, and PT9. Clearly, considering the higher-order motion parameters b and c, the impulse responses of targets PT1, PT5, and PT9 with different range and azimuth positions are visibly well focused by the proposed method. However, neglecting the motion parameters b and c, the impulse responses of targets PT1, PT5, and PT9 using the proposed method have deterioration with different degrees. The neglected parameter c can degrade the near-sidelobe levels with asymmetry distortions, which means that there are deteriorations in the peak sidelobe ratio (PSLR) and integrated sidelobe ratio (ISLR), and the neglected parameter b can degrade both the 3 dB width of the main lobe (i.e., the resolution) and the near-sidelobe levels, as shown in Figure 12. The impulse responses of targets PT1 and PT9 on the scene edges using

FDA have deteriorations. The neglected azimuth and cross-coupling spatial variations brought by acceleration are the main causes of the problem and they increase with resolutions and scene sizes.

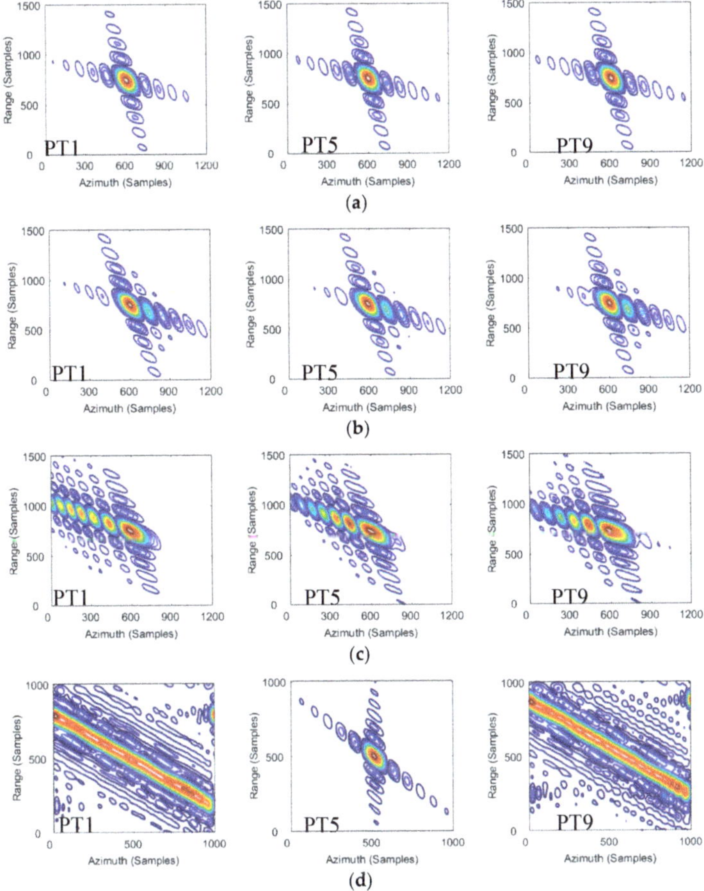

Figure 12. Comparative results of target PT1, PT5, and PT9. (**a**) Proposed method, (**b**) c not considered, (**c**) b not considered, and (**d**) FDA.

To quantify the precision of the proposed method, IRW, PSLR, and ISLR are used as performance measures. The results are listed in Table 2. Both the contour plots and image quality parameters demonstrate the effectiveness of the proposed method.

Table 2. Image quality parameters.

Method	Target	Range			Azimuth		
		IRW (m)	PSLR (dB)	ISLR (dB)	IRW (m)	PSLR (dB)	ISLR (dB)
Proposed	PT1	0.266	−13.21	−9.99	0.247	−13.17	−9.92
	PT5	0.265	13.23	10.01	0.240	−13.22	−10.03
	PT9	0.266	−13.19	−9.98	0.241	−13.15	−9.95
FDA	PT1	0.266	−13.24	−9.96	0.893	−6.02	−4.49
	PT5	0.266	−13.25	−10.09	0.242	−13.23	−10.07
	PT9	0.267	−13.17	−10.01	1.302	−4.74	−3.56

Case 2: In this case, a, b, and c are set to larger values compared with those in Case 1, $a = (3.2, 4.1, -2.7)$ m/s^2, $b = (0.32, -0.56, -0.17)$ m/s^3, and $c = (-0.032, -0.037, 0.024)$ m/s^4. The ideal azimuth resolution is 0.364 m, the height of target PT9 is set to 300 m with respect to the reference target PT5, and other simulation parameters are listed in Table 1.

By performing the focusing method, simulation result of targets PT1, PT5, and PT9 is shown in Figure 13. The wavenumber domain algorithm [17] and BPA [29] are used for comparisons. Clearly, the results using the BPA and proposed method are visibly well focused, whereas the results on the edges achieved by [17] are not because the spatial variations introduced by the motion parameters a, b, and c are not considered. Moreover, the target height of PT9 leads to a greater deterioration on the focused result compared with that of the flat one.

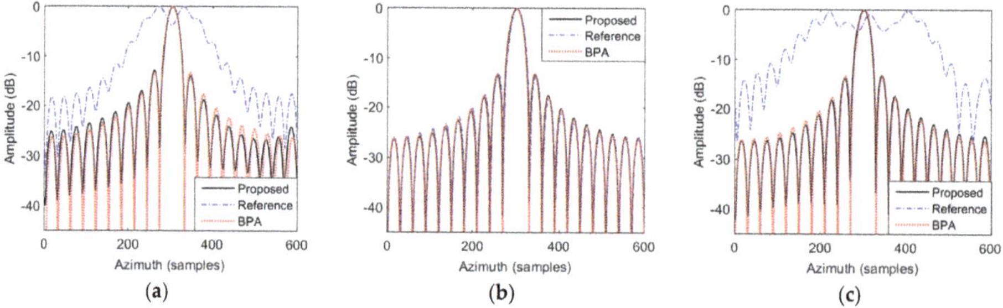

Figure 13. Comparative results of azimuth point impulse responses processed by proposed method [17], and BPA. (**a**) Target PT1. (**b**) Target PT5. (**c**) Target PT9.

The quality parameters of azimuth point impulse responses are listed in Table 3. It is worth noting that the quality parameters of the proposed method are close to those of the BPA. In particular, the computational load of the proposed approach is much lower than that of the BPA. All these indicate that the proposed method can be well applied to the HRHS SAR with maneuvers.

Table 3. Image quality parameters.

Method	Target	IRW (m)	PSLR (dB)	ISLR (dB)
	PT1	0.367	−13.16	−9.87
Proposed	PT5	0.365	−13.21	−10.02
	PT9	0.362	−13.14	−9.94
	PT1	1.031	−6.11	−4.56
[17]	PT5	0.364	−13.24	−10.06
	PT9	1.135	−5.78	−4.12
	PT1	0.365	−13.24	−10.04
BPA	PT5	0.364	−13.27	−10.08
	PT9	0.361	−13.25	−10.05

5.2. Experiment 2

In the following, a comparison of the proposed approach and [17] is made. Since the highly-squinted SAR data with maneuvers are not available, a HRHS airborne SAR raw signal simulation through time domain echo generation method is performed in this subsection. The data set contains curved flight path. The carrier frequency is 35 GHz, the bandwidth of the transmitted signal is 400 MHz, reference range is 26 km, and squint angle is 63°. The scene size in range and azimuth directions are respectively 1.4 km and 1 km, and the azimuth resolution is 0.428 m. The motion parameters—namely, the velocity v, acceleration a, and higher-order motion parameters b and c are listed in Table 4. The data are focused by using the proposed approach. Moreover, a comparative focusing result of [17] is provided to demonstrate the superiority of the proposed method.

Table 4. Motion parameter settings.

Motion Parameter	Value
Velocity v	$(0, 110, -10)$ m/s
Acceleration a	$(1.21, 1.43, -0.74)$ m/s^2
3rd-order Paramater b	$(-0.1, 0.2, 0.2)$ m/s^3
4th-order Paramater c	$(0.007, -0.015, 0.004)$ m/s^4

Figure 14 shows the comparative results. Clearly, the entire scene is well focused by the proposed approach, including the edge regions, as shown in Figure 14a. However, the imaging result in Figure 14b has a great deterioration in the edge regions, which are noted from the zoom-in version of the dot-line rectangle area. It is because [17] ignores the spatial variations introduced by a, b, and c.

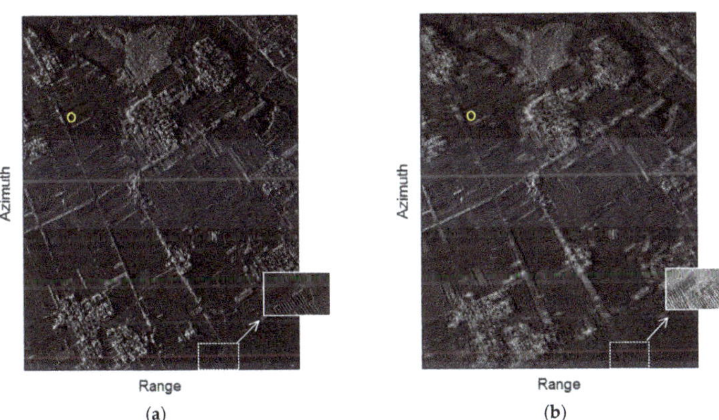

Figure 14. Imaging results (a) by proposed approach and (b) [17].

Figure 15 shows the zoom-in version of TFDs of the highlighted elliptic areas in Figure 14. The TFDs of the proposed method have a good energy aggregation, however, the TFDs of [17] have great energy dispersion. Moreover, the time-frequency resolution (TFR) of the proposed approach is higher than that of [17], which is seen from the dot-line rectangle area. It is also observed that the TFDs of the proposed approach are vertical curves while that of [17] have slight slopes. The azimuth profiles of the point in the highlighted elliptic areas are shown in Figure 16. It is evident that serious distortion and smearing occur in [17], while the proposed method provides well-focused performance. According to the above analyses, it is concluded that the proposed approach can perform well in HRHS airborne SAR with maneuvers.

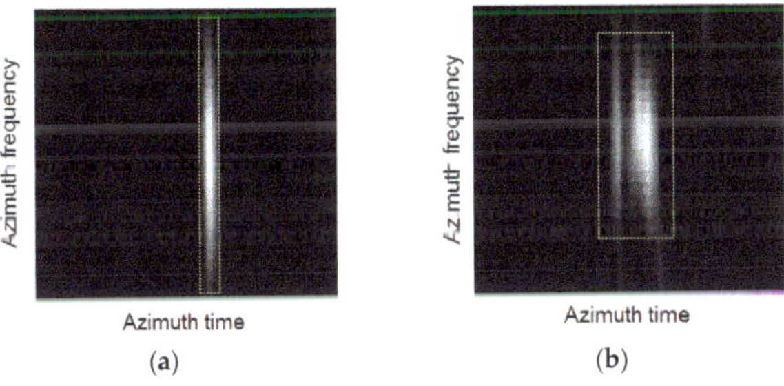

Figure 15. TFDs of rectangular domain in Figure 14. (a) Using the proposed approach. (b) Using [17].

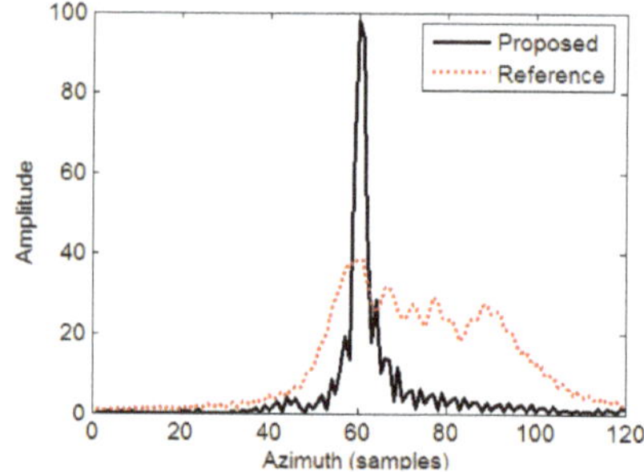

Figure 16. Comparison of azimuth pulse response.

6. Conclusions

SAR has been widely applied for remote sensing. However, the problems caused by the maneuvers affect the performance of traditional focusing method for the HRHS cases. In this paper, a FMPM is introduced to describe the curved path. Considering the third- and higher-order motion parameters, our analyses indicate that the spatial variations in range, azimuth, and height directions will severely impair the image quality if they are not properly accounted for during the processing. To solve this problem, we have developed a polynomial phase filter to remove the cross-coupling variations and a Stolt mapping function to the range and azimuth terms. The proposed approach is efficient, easy to implement, and can process the HRHS SAR data with maneuvers. Moreover, implementation considerations are provided. Validity and applicability are studied through theoretical analyses and numerical experiments.

Author Contributions: S.T. conceived the main idea; L.Z. and H.C.S. conceived and designed the experiments; S.T. and H.C.S. analyzed the data and wrote the paper.

Acknowledgments: This work was supported in part by the National Natural Science Foundation of China under grants 61601343 and 61671361, in part by China Postdoctoral Science Foundation Funded Project under grant 2016M600768, in part by the National Defense Foundation of China, and in part by the Fund for Foreign Scholars in University Research and Teaching Programs (the 111 project) under grant No. B18039.

Conflicts of Interest: The authors declare no conflict of interest.

Appendix A

In order to obtain the 2-D spectrum of Equation (11), we use the POSP and have

$$\sum_{n=1}^{4} \frac{1}{(n-1)!}\left(\mu_n - \mu_n^{ref}\right)\eta^{n-1} + \sum_{n=2}^{4} \frac{1}{(n-1)!}\chi_n\eta^{n-1} = -f_k \tag{29}$$

where $f_k = c f_\eta / 2(f_c + f_r)$. By using MSR [39–41], the stationary point η^* is derived as

$$\eta^* = -g_1\left(f_k + \mu_1 - \mu_1^{ref}\right) + g_2\left(f_k + \mu_1 - \mu_1^{ref}\right)^2 - g_3\left(f_k + \mu_1 - \mu_1^{ref}\right)^3 \tag{30}$$

where the coefficients are

$$g_1 = \frac{1}{\mu_2 - \mu_2^{ref} + \chi_2} \tag{31}$$

$$g_2 = \frac{\mu_3 - \mu_3^{ref} + \chi_3}{2\left(\mu_2 - \mu_2^{ref} + \chi_2\right)^3} \tag{32}$$

$$g_3 = \frac{\left(\mu_3 - \mu_3^{ref} + \chi_3\right)^2}{2\left(\mu_2 - \mu_2^{ref} + \chi_2\right)^5} - \frac{\left(\mu_4 - \mu_4^{ref} + \chi_4\right)}{6\left(\mu_2 - \mu_2^{ref} + \chi_2\right)^4} \tag{33}$$

According to Equation (30), the 2-D spectrum is derived as

$$S_2\left(f_r, f_\eta\right) \approx \varepsilon_0 \omega_r(f_r)\omega_a\left(f_\eta\right)\exp\left\{j\pi \frac{4(f_r + f_c)}{c}\left(k_0 + k_1 f_k + k_2 f_k^2 + k_3 f_k^3 + k_4 f_k^4\right)\right\} \tag{34}$$

where $k_0 = \sum\limits_{n=0}^{4}\beta_n\left(\mu_1 - \mu_1^{ref}\right)^n$, $k_1 = \sum\limits_{n=1}^{4}n\beta_n\left(\mu_1 - \mu_1^{ref}\right)^{n-1}$, $k_2 = 0.5\sum\limits_{n=2}^{4}n(n-1)\beta_n\left(\mu_1 - \mu_1^{ref}\right)^{n-2}$,

$k_3 = \sum\limits_{n=3}^{4}4^{n-3}\beta_n\left(\mu_1 - \mu_1^{ref}\right)^{n-3}$, and $k_4 = \beta_4$, with β_n being the Doppler parameters of the range

history, all of which are spatially variant terms. They are calculated as $\beta_0 = -\left(\mu_0 - \mu_0^{ref}\right)$, $\beta_1 = 0$,
$\beta_2 = g_1/2$, $\beta_3 = -g_2/3$, and $\beta_4 = g_3/4$.

Appendix B

For the second-order coefficient k_2 in Equation (34), we expand β_2, β_3, and β_4 using the Taylor series

$$\beta_2 = \beta_2^{ref} + a_1\left(\mu_1 - \mu_1^{ref}\right) + a_2\left(\mu_1 - \mu_1^{ref}\right)^2 + a_3\left(k_0 - k_0^{ref}\right) \tag{35}$$

$$\beta_3 = \beta_3^{ref} + a_4\left(\mu_1 - \mu_1^{ref}\right) + a_5\left(\rho_1 - \rho_1^{ref}\right) \tag{36}$$

$$\beta_4 = \beta_4^{ref} \tag{37}$$

where $\rho_1 = k_1/\left(\mu_1 - \mu_1^{ref}\right)$, $k_0^{ref} = 0$, and $\rho_1^{ref} = 2\beta_2^{ref}$. $a_i(i = 1, \cdots, 5)$ are the Taylor expansion coefficients. Substituting Equations (35)–(37) into k_2 leads to

$$\begin{aligned}
k_2 = {} & p_2(\chi_2, \chi_3, \chi_4) \cdot k_0 + q_2(\chi_2, \chi_3, \chi_4) \cdot k_1 \\
& + l_2(\chi_2, \chi_3, \chi_4) \cdot \left(\mu_1 - \mu_1^{ref}\right) \\
& + l_3(\chi_2, \chi_3, \chi_4) \cdot \left(\mu_1 - \mu_1^{ref}\right)^2 \\
& + z_2(\chi_2, \chi_3, \chi_4)
\end{aligned} \tag{38}$$

where $p_2(\chi_2, \chi_3, \chi_4) = a_3$, $q_2(\chi_2, \chi_3, \chi_4) = 3a_5$, $l_2(\chi_2, \chi_3, \chi_4) = a_1 + 3\beta_3^{ref} - 6a_5\beta_2^{ref}$, $l_3(\chi_2, \chi_3, \chi_4) = a_2 + 3a_4 + 6\beta_4^{ref}$, and $z_2(\chi_2, \chi_3, \chi_4) = \beta_2^{ref}$.

For the third-order coefficient k_3 in Equation (34), β_3 is re-expanded using Taylor series with different variables as those of Equation (36)

$$\beta_3 = \beta_3^{ref} + b_1\left(\mu_1 - \mu_1^{ref}\right) + b_2\left(k_0 - k_0^{ref}\right) \tag{39}$$

where $b_i(i = 1, 2)$ are the Taylor expansion coefficients. Substituting Equation (39) into k_3, we obtain

$$\begin{aligned}
k_3 = {} & p_3(\chi_2, \chi_3, \chi_4) \cdot k_0 + q_3(\chi_2, \chi_3, \chi_4) \cdot k_1 \\
& + l_4(\chi_2, \chi_3, \chi_4) \cdot \left(\mu_1 - \mu_1^{ref}\right) + z_3(\chi_2, \chi_3, \chi_4)
\end{aligned} \tag{40}$$

where $p_3(\chi_2, \chi_3, \chi_4) = b_2$, $q_3(\chi_2, \chi_3, \chi_4) = 0$, $l_4(\chi_2, \chi_3, \chi_4) = b_1 + 4\beta_4^{ref}$, and $z_3(\chi_2, \chi_3, \chi_4) = \beta_3^{ref}$.

For the fourth-order coefficient k_4 in Equation (34), we substitute Equation (37) into k_4 to yield

$$k_4 = p_4(\chi_2, \chi_3, \chi_4) \cdot k_0 + q_4(\chi_2, \chi_3, \chi_4) \cdot k_1 + z_4(\chi_2, \chi_3, \chi_4) \tag{41}$$

where $p_4(\chi_2, \chi_3, \chi_4) = 0$, $q_4(\chi_2, \chi_3, \chi_4) = 0$, and $z_4(\chi_2, \chi_3, \chi_4) = \beta_4^{ref}$.

Substituting Equations (38), (40), and (41) into Equation (34), we obtain the 2-D spectrum in (13). According to phase decomposition, the equations in Equation (18) are rewritten as

$$
\begin{cases}
a_1 + 3\beta_3^{ref} - 6a_5\beta_2^{ref} = 0 \\
a_2 + 3a_4 + 6\beta_4^{ref} = 0 \\
b_1 + 4\beta_4^{ref} = 0
\end{cases}
\tag{42}
$$

Their solutions are

$$
\chi_2 = -\left(C_3 + C_1^2\right)/C_2
\tag{43}
$$

$$
\chi_3 = -\left(C_3 C_1 + C_1^3\right)/C_2
\tag{44}
$$

$$
\chi_4 = C_3^2 - C_3 C_1^2 / C_2
\tag{45}
$$

where $C_1 = \partial\mu_2/\partial\mu_1|_{ref}$, $C_2 = \partial^2\mu_2/\partial\mu_1^2|_{ref}$, and $C_3 = \partial\mu_3/\partial\mu_1|_{ref}$.

References

1. Cumming, I.G.; Wong, F.H. *Digital Processing of Synthetic Aperture Radar Data: Algorithms and Implementation*; Artech House: Norwood, MA, USA, 2005.
2. Sun, G.-C.; Xing, M.; Xia, X.; Yang, J.; Wu, Y.; Bao, Z. A Unified Focusing Algorithm for Several Modes of SAR Based on FrFT. *IEEE Trans. Geosci. Remote Sens.* **2013**, *51*, 3139–3155. [CrossRef]
3. Peng, X.; Wang, Y.; Hong, W.; Wu, Y. Autonomous Navigation Airborne Forward-Looking SAR High Precision Imaging with Combination of Pseudo-Polar Formatting and Overlapped Sub-Aperture Algorithm. *Remote Sens.* **2013**, *5*, 6063–6078. [CrossRef]
4. Stroppiana, D.; Azar, R.; Calò, F.; Pepe, A.; Imperatore, P.; Boschetti, M.; Silva, J.M.N.; Brivio, P.A.; Lanari, R. Integration of Optical and SAR Data for Burned Area Mapping in Mediterranean Regions. *Remote Sens.* **2015**, *7*, 1320–1345. [CrossRef]
5. Carrara, W.G.; Goodman, R.S.; Majewski, R.M. *Spotlight Synthetic Aperture Radar: Signal Processing Algorithms*; Artech House: Norwood, MA, USA, 1995.
6. Ausherman, D.A.; Kozma, A.; Walker, J.L.; Jones, H.M.; Poggio, E.C. Developments in Radar Imaging. *IEEE Trans. Aerosp. Electron. Syst.* **1984**, *AES-20*, 363–400. [CrossRef]
7. Farrell, J.L.; Mims, J.H.; Sorrell, A. Effects of Navigation Errors in Maneuvering SAR. *IEEE Trans. Aerosp. Electron. Syst.* **1973**, *AES-9*, 758–776. [CrossRef]
8. Robinson, P.N. Depth of Field for SAR with Aircraft Acceleration. *IEEE Trans. Aerosp. Electron. Syst.* **1984**, *AES-20*, 603–616. [CrossRef]
9. Axelsson, S.R.J. Mapping Performance of Curved-path SAR. *IEEE Trans. Geosci. Remote Sens.* **2002**, *40*, 2224–2228. [CrossRef]
10. Zhou, P.; Xing, M.; Xiong, T.; Wang, Y.; Zhang, L. A Variable-decoupling- and MSR-based Imaging Algorithm for a SAR of Curvilinear Orbit. *IEEE Geosci. Remote Sens. Lett.* **2011**, *8*, 1145–1149. [CrossRef]
11. Eldhuset, K. A New Fourth-order Processing Algorithm for Spaceborne SAR. *IEEE Trans. Aerosp. Electron. Syst.* **1998**, *34*, 824–835. [CrossRef]
12. Eldhuset, K. Ultra High Resolution Spaceborne SAR Processing with EETE4. In Proceedings of the IGARSS 2011, Vancouver, BC, Canada, 24–29 July 2011; pp. 2689–2691.
13. Luo, Y.; Zhao, B.; Han, X.; Wang, R.; Song, H.; Deng, Y. A Novel High-order Range Model and Imaging Approach for High-resolution LEO SAR. *IEEE Trans. Geosci. Remote Sens.* **2014**, *52*, 3473–3485. [CrossRef]
14. Wang, P.; Liu, W.; Chen, J.; Niu, M.; Yang, W. A High-order Imaging Algorithm for High-resolution Spaceborne SAR Based on a Modified Equivalent Squint Range Model. *IEEE Trans. Geosci. Remote Sens.* **2015**, *53*, 1225–1235. [CrossRef]
15. Huang, L.J.; Qiu, X.L.; Hu, D.H.; Han, B.; Ding, C.B. Medium-earth-orbit SAR Focusing using Range Doppler Algorithm with Integrated Two-step Azimuth Perturbation. *IEEE Geosci. Remote Sens. Lett.* **2015**, *12*, 626–630. [CrossRef]

16. Tang, S.; Zhang, L.; Guo, P.; Liu, G.; Zhang, Y.; Li, Q.; Gu, Y.; Lin, C. Processing of Monostatic SAR Data with General Configurations. *IEEE Trans. Geosci. Remote Sens.* **2015**, *53*, 6529–6546. [CrossRef]
17. Li, Z.; Xing, M.; Xing, W.; Liang, Y.; Gao, Y.; Dai, B.; Hu, L.; Bao, Z. A Modified Equivalent Range Model and Wavenumber-domain Imaging Approach for High-resolution-high-squint SAR with Curved Trajectory. *IEEE Trans. Geosci. Remote Sens.* **2017**, *55*, 3721–3734. [CrossRef]
18. Xing, M.; Jiang, X.; Wu, R.; Zhou, F.; Bao, Z. Motion Compensation for UAV SAR Based on Raw Radar Data. *IEEE Trans. Geosci. Remote Sens.* **2009**, *47*, 2870–2883. [CrossRef]
19. Smith, A.M. A New Approach to Range-Doppler SAR Processing. *Int. J. Remote Sens.* **1991**, *12*, 235–251. [CrossRef]
20. Raney, R.K.; Runge, H.; Bamler, R.; Cumming, I.G.; Wong, F.H. Precision SAR Processing using Chirp Scaling. *IEEE Trans. Geosci. Remote Sens.* **1994**, *32*, 786–799. [CrossRef]
21. Cafforio, C.; Prati, C.; Rocca, F. SAR Data Focusing using Seismic Migration Techniques. *IEEE Trans. Aerosp. Electron. Syst.* **1991**, *27*, 194–207. [CrossRef]
22. Tang, S.; Zhang, L.; Guo, P.; Liu, G.; Sun, G.-C. Acceleration Model Analyses and Imaging Algorithm for Highly Squinted Airborne Spotlight Mode SAR with Maneuvers. *IEEE J. Sel. Top. Appl. Earth Obs. Remote Sens.* **2015**, *8*, 1120–1131. [CrossRef]
23. Li, Z.; Xing, M.; Liang, Y.; Gao, Y.; Chen, J.; Huai, Y.; Zeng, L.; Sun, G.-C.; Bao, Z. A Frequency-domain Imaging Algorithm for Highly Squinted SAR Mounted on Maneuvering Platforms with Nonlinear Trajectory. *IEEE Trans. Geosci. Remote Sens.* **2016**, *54*, 4023–4038. [CrossRef]
24. Liu, G.; Li, P.; Tang, S.; Zhang, L. Focusing Highly Squinted Data with Motion Errors Based on Modified Non-linear Chirp scaling. *IET Radar Sonar Navig.* **2013**, *7*, 568–578.
25. Wu, J.; Xu, Y.; Zhong, X.; Yang, J.M. A Three-Dimensional Localization Method for Multistatic SAR Based on Numerical Range-Doppler Algorithm and Entropy Minimization. *Remote Sens.* **2017**, *9*, 470. [CrossRef]
26. Tang, S.; Lin, C.; Zhou, Y.; So, H.C.; Zhang, L.; Liu, Z. Processing of Long Integration Time Spaceborne SAR Data with Curved Orbit. *IEEE Trans. Geosci. Remote Sens.* **2018**, *56*, 888–904. [CrossRef]
27. Wu, C.; Liu, K.Y.; Jin, M. Modeling and a Correlation Algorithm for Spaceborne SAR Signals. *IEEE Trans. Aerosp. Electron. Syst.* **1982**, *AES-18*, 563–575. [CrossRef]
28. Yang, L.; Zhou, S.; Zhao, L.; Xing, M. Coherent Auto-Calibration of APE and NsRCM under Fast Back-Projection Image Formation for Airborne SAR Imaging in Highly-Squint Angle. *Remote Sens.* **2018**, *10*, 321. [CrossRef]
29. Munson, D.; O'Brien, J.; Jenkins, W. A Tomographic Formulation of Spotlight-mode Synthetic Aperture Radar. *Proc. IEEE* **1983**, *71*, 917–925. [CrossRef]
30. Ulander, L.M.H.; Hellsten, H.; Stenström, G. Synthetic Aperture Radar Processing using Fast Factorized Back-projection. *IEEE Trans. Aerosp. Electron. Syst.* **2003**, *39*, 760–776. [CrossRef]
31. Zhang, L.; Li, H.-L.; Qiao, Z.-J.; Xu, Z.-W. A Fast BP Algorithm with Wavenumber Spectrum Fusion for High-resolution Spotlight SAR Imaging. *IEEE Geosci. Remote Sens. Lett.* **2014**, *11*, 1460–1464. [CrossRef]
32. Jokowartz, C.V.; Wahl, D.E.; Eichel, P.H.; Ghiglia, D.C.; Thompson, P.A. *Spotlight-Mode Synthetic Aperture Radar: A Signal Processing Approach*; Kluwer: Norwell, MA, USA, 1996.
33. Taylor, A.E.; Mann, W.R. *Advanced Calculus*, 2nd ed.; John Wiley & Sons: New York, NY, USA, 1972.
34. Jankech, A. Using Vluster Multifunctions for Fecomposition Theorems. *Int. J. Pure Appl. Math.* **2008**, *46*, 303–312.
35. Wong, F.H.; Yco, T.S. New Application of Nonlinear Chirp Scaling in SAR Data Processing. *IEEE Trans. Geosci. Remote Sens.* **2001**, *39*, 946–953. [CrossRef]
36. An, D.; Huang, X.; Jin, T.; Zhou, Z. Extended Nonlinear Chirp Scaling Algorithm for High-resolution Highly Squint SAR Data Focusing. *IEEE Trans. Geosci. Remote Sens.* **2012**, *50*, 3595–3609. [CrossRef]
37. Xu, G.; Xing, M.; Zhang, L.; Bao, Z. Robust Autofocusing Approach for Highly Squinted SAR Imagery using the Extended Wavenumber Algorithm. *IEEE Trans. Geosci. Remote Sens.* **2013**, *51*, 5031–5046. [CrossRef]
38. Zhang, L.; Sheng, J.; Xing, M.; Qiao, Z.; Xiong, T.; Bao, Z. Wavenumber-domain Autofocusing for Highly Squinted UAV SAR Imagery. *IEEE Sens. J.* **2012**, *12*, 1574–1588. [CrossRef]
39. Neo, Y.L.; Wong, F.; Cumming, I.G. A Two-dimensional Spectrum for Bistatic SAR Processing using Series Reversion. *IEEE Geosci. Remote Sens. Lett.* **2007**, *4*, 93–96. [CrossRef]

40. Neo, Y.L.; Wong, F.; Cumming, I.G. Processing of Azimuth Invariant Bistatic SAR Data using the Range Doppler Algorithm. *IEEE Trans. Geosci. Remote Sens.* **2008**, *46*, 14–21. [CrossRef]
41. Hu, C.; Liu, Z.; Long, T. An Improved CS Algorithm Based on the Curved Trajectory in Geosynchronous SAR. *IEEE J. STARS* **2012**, *5*, 795–808. [CrossRef]

 *remote sensing*

Article

Focusing High-Resolution Airborne SAR with Topography Variations Using an Extended BPA Based on a Time/Frequency Rotation Principle

Chunhui Lin [1], Shiyang Tang [1,]*, Linrang Zhang [1] and Ping Guo [2]

[1] National Lab of Radar Signal Processing, Xidian University, Xi'an 710071, China; larrylinch@163.com (C.L.); lrzhang@xidian.edu.cn (L.Z.)

[2] College of Communication and Information Engineering, Xi'an University of Science and Technology, Xi'an 710054, China; guoping@xust.edu.cn

* Correspondence: sytang@xidian.edu.cn; Tel.: +86-153-1992-2377

Received: 4 July 2018; Accepted: 10 August 2018; Published: 13 August 2018

Abstract: With the increasing requirement for resolution, the negligence of topography variations causes serious phase errors, which leads to the degradation of the focusing quality of the synthetic aperture (SAR) imagery, and geometric distortion. Hence, a precise and fast algorithm is necessary for high-resolution airborne SAR. In this paper, an extended back-projection (EBP) algorithm is proposed to compensate the phase errors caused by topography variations. Three-dimensional (3D) variation will be processed in the time-domain for high-resolution airborne SAR. Firstly, the quadratic phase error (QPE) brought by topography variations is analyzed in detail for high-resolution airborne SAR. Then, the key operation, a time-frequency rotation operation, is applied to decrease the samplings in the azimuth time-domain. Just like the time-frequency rotation of the conventional two-step approach, this key operation can compress data in an azimuth time-domain and it reduces the computational burden of the conventional back-projection algorithm, which is applied lastly in the time-domain processing. The results of the simulations validate that the proposed algorithm, including frequency-domain processing and time-domain processing can obtain good focusing performance. At the same time, it has strong practicability with a small amount of computation, compared with the conventional algorithm.

Keywords: frequency-domain processing; extended back-projection algorithm; topography variations; computational burden; high resolution

---

1. Introduction

Over the past few decades, airborne synthetic aperture radar (SAR) has been widely adopted in remote sensing areas to produce a two-dimensional (2-D) high-resolution microwave imagery of observed scenes under all-weather and all-day conditions [1–3]. The SAR technology is important in object detection, localization, and deformation monitoring [4–6]. A variety of airborne SAR systems have been applied in practical applications, such as PAMIR, CARABAS-II, and unmanned aerial vehicle (UAV) SAR. The airborne SAR systems nowadays can illuminate the observed area at different angles, based on the SAR usually being divided into side-board SAR and squint SAR [7]. As an effective sensor, high squint SAR can also provide information about the surface structure, and can increase the flexibility to obtain the imagery of the desired district within a single pass of the airplane. Many efficient algorithms have been proposed to resolve the problems, i.e., range cell migration, azimuth spectrum aliasing, and 2-D spatial-variant Doppler frequency modulation rate [8–10]. The topography of observed scenes is also usually neglected in these conventional airborne SAR algorithms, which means that the ground is assumed to be a flat surface [11–13]. However, the assumption of these algorithms will be invalid, especially with increasing resolution nowadays [14]. The negligence of topography variations will cause two main problems: degradation of the focusing quality of imagery, and geometric distortion. Therefore, to satisfy the increasing demand for the high-resolution of airborne SAR systems (such as millimeter-wave SAR systems), a precise focusing algorithm for high-resolution highly squinted airborne SAR is unquestionably necessary to solve problems that are caused by topography variations.

SAR image formation for high-resolution SAR basically consists of a phase-corrected integration of the raw data samples. The different processing algorithms deviate from each other in terms of how accurately and efficiently they implement the spatially adaptive summation. In principle, the algorithms can be mainly divided into two categories: frequency-domain [15–24] and time-domain algorithms [25–28]. The former ones are aimed at raising the efficiency of the processor, but they have the limitations of integration time, flight track, and topography, which restricts the application of frequency-domain algorithms. The most general frequency-domain algorithm is the range Doppler algorithm (RDA), which decouples the range and azimuth processing with the sacrifice of focusing accuracy. Frequency-domain algorithms also include the chirp scaling (CS) algorithm [15], the Omega-k algorithm [16], and their extended forms for squinted airborne SAR. They are usually based on the assumption of a strictly straight trajectory. The nonlinear chirp scaling (NLCS) algorithm [17] and its extension [18–22] are the common algorithms that are used to deal with the spatial variance, especially the azimuth variance. These frequency-domain algorithms are developed by scholars before the graphics processing unit was applied. In general, these aforementioned algorithms have not taken the topography into account. The precise topography and aperture dependent (PTA) algorithms [25] are proposed to realize the motion compensation based on block processing. To avoid discontinuities at block borders, it usually takes a 50% overlap, which results in a relatively high computational burden. The subaperture topography- and aperture- dependent (SATA) algorithm [26] is another typical algorithm that increases the accuracy of the topography variants. This uses the short time Fourier transform in subaperture along the azimuthal direction.

The typical time-domain algorithm is the time-domain back-projection (BP) algorithm, which has been widely accepted as a precise algorithm for all SAR reconstructions [29,30]. In principle, BP algorithm is a process of pulse-by-pulse and pixel-by-pixel. Thus, it can be applied in almost every SAR mode, configuration, and terrain. At the same time, the digital elevation model (DEM) has been a mature technology in the remote sensing field [31–33]. The combination of the BP algorithm and DEM can perfectly realize the topography-dependent motion compensation. Due to the pixel-by-pixel character, the biggest disadvantage of the BP algorithm is the computational burden. Usually, the computation of the time-domain BP algorithm is proportional for a desired image with, and azimuth sampling points. Compared with frequency-domain algorithms, the time-domain BP algorithm provides lower adaptability and practicability. To improve the computational efficiency for time-domain algorithms, two major approaches have been developed: parallel computing platforms and incremental modifications of the BP algorithm. Many methods have been proposed to improve the operational efficiency of the BP algorithm. The representative ones are the fast back projection (FBP) algorithm [34] and the fast factorized back projection (FFBP) algorithm [35]. The FBP algorithm and FFBP algorithm both consist of a number of processing stages, and combine the subaperture technology with polar grids processing. To a certain degree, they both reduce the computational burden, while keep most of the advantages of the original BP algorithm. The extensions of the FBP algorithm and the FFBP algorithm were applied in different SAR scenes [36–43]. However, the subaperture will introduce problems of subaperture division and sub-image mosaic [44]. Also, topography variations are taken in account, which will cause image distortion. In general, for high-resolution airborne SAR systems, it becomes very difficult for the methods mentioned above, to realize the focusing of the observed scene, with the increase in resolution, squint angle, and scene coverage. A fast focus algorithm based on full-aperture processing is necessary, especially when topography variations cannot be ignored.

However, some overlap between the blocks is considered, to reduce the appearance of sidelobe in the SATA algorithm.

In this paper, we proposed an extended back-projection (EBP) algorithm for the squinted high-resolution airborne SAR, with the consideration of topography variations. To avoid subaperture division and image mosaic, the proposed algorithm will be based on full-aperture processing, rather than partitioned processing. The range domain data or azimuth domain data is processed as a bulk. This proposed algorithm applies a time-frequency rotation operation, which is similar to the two-steps approach in certain extent. Compared with the conventional time-domain BP algorithm, the proposed EBP algorithm needs lesser amounts of computation. At the same time, it remains high-resolution performance for high-resolution airborne SAR systems with topography variations.

The paper is organized as follows. Section 2 is dedicated to the model of high-resolution airborne SAR with topography variations and analysis of phase errors without consideration, and the derivation of corresponding expressions are given. The details of the proposed EBP algorithm, i.e., the frequency-domain processing and time-domain processing, are explained in Section 3. Then, several simulation results of the performance of the proposed algorithm are provided in Section 4. Section 5 discusses the problems about parameter selection and the computational burden for the proposed EBP algorithm. The final conclusion is given in Section 6.

2. Modeling and Analysis

2.1. Modeling of Airborne SAR

Traditionally, the movement of airborne SAR is assumed as a linear uniform motion in a straight line, with the ground scene being flat. It was reasonable for the short integration time, low-resolution airborne SAR. The hyperbolic range model, neglecting the impacts of topography variations on the imaging results, would introduce significant phase errors with the increasing resolution for airborne SAR systems. During the acquisition interval, neither the motion error effects, nor the topography variations could be ignored. A general geometry of high-resolution airborne SAR is shown in Figure 1.

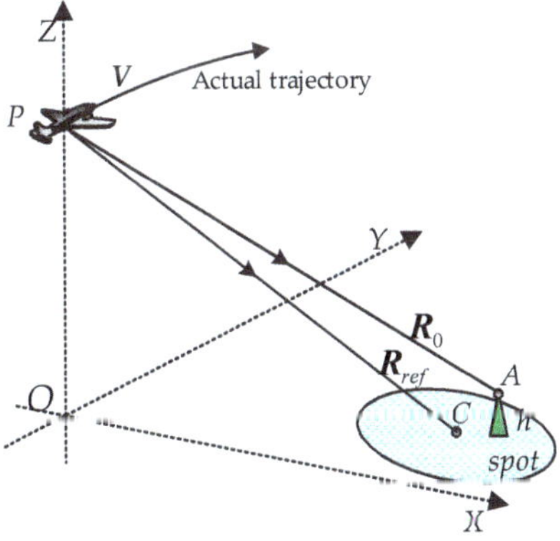

Figure 1. Geometrical model of high-resolution airborne synthetic aperture (SAR).

It is noted that point P is the reference position of the fight path at the aperture center moment (ACM), and points C and A are respectively the reference and arbitrary targets on the ground scene, h is the height of the point A, R_{ref} and R_0 are respectively the slant range vectors from P to C, and P to A at ACM, and V is the velocity vector of the platform. With reference to the imaging geometry, the slant range history of point A is expressed as:

$$r(\eta; A) = r_i(\eta; A) + r_e(\eta; A) + r_t(\eta; A) \tag{1}$$

where η is the slow time, $r_i(\eta; A)$, $r_e(\eta; A)$, and $r_t(\eta; A)$ are respectively the range history, motion errors, and the topography variations of target A, and $r_i(\eta; A)$ are expressed as:

$$r_i(\eta; A) = |\boldsymbol{R}_0 - \boldsymbol{V} \cdot \eta| \tag{2}$$

where $|\cdot|$ denotes the norm operation. In this general model, $r_e(\eta; A)$ and $r_t(\eta; A)$ cannot be ignored in the case of high-resolution.

According to Equation (1), the received echo of point A is expressed as:

$$
\begin{aligned}
S_0(t_r, \eta) = \ & \varepsilon_0 \omega_r(t_r - 2r(\eta; A)/c)\omega_a(\eta - \eta_0) \\
& \cdot \exp(-j4\pi f_c r(\eta; A)/c) \cdot \exp\left(j\pi\gamma[t_r - 2r(\eta; A)/c]^2\right)
\end{aligned}
\tag{3}
$$

where t_r is the range fast time, f_c and γ are the carrier frequency and the frequency modulation (FM) rate of the transmitted signal, respectively, ε_0 is the complex scattering coefficient, $\omega_r(\cdot)$ and $\omega_a(\cdot)$ are respectively the range and azimuth envelopes in the time-domain, and η_0 denotes the Doppler center time of the point target [45,46].

2.2. Problem

In conventional SAR imaging processing, the phase error brought about by topography variation was not considered, which would greatly deteriorate the final image and limit the ground scene size. In this subsection, the quadratic phase error (QPE) brought by topography variation $r_t(\eta; A)$ was analyzed for high-resolution airborne SAR, according to Equations (1) and (2). Utilizing the parameters listed in Table 1, Figure 2 shows the absolute QPEs with different variables, i.e., resolution, squint angle, and scene size. The unit of the figures was π. Figure 2a,b is respectively, the simulation results in the range/height and azimuth/height domains with azimuth resolution and the squinted angle, respectively being 0.25 m and 60°. Figure 2c,d is the simulation results in the resolution/height and squint angle/height domains, respectively. Clearly, the maximum QPE in each domain was far larger than $\pi/4$, which could not be accepted in the high-resolution airborne SAR imaging. Moreover, the QPE increased with the scene coverage, azimuth resolutions, and the squint angles, as shown in Figure 2. It was concluded that the effects that were brought by topography variations cannot be ignored in the case of high-resolution, large-swath, and high squint angles. Figure 3 shows the absolute position errors (PEs) that were introduced by topography variations. The unit of the figures was impulse-response width (IRW). It was seen that the PEs were all generally larger than one IRW in each height cell, and that they would greatly distort the final focused image.

Table 1. Parameter settings.

Parameters	Value
Carrier frequency	9.65 GHz
Pulse duration	2.5 µs
Pulse Bandwidth	300 MHz
Sampling frequency	420 MHz
Reference slant range	25.0 km
Altitude	8.0 km
Velocity vector	(0, 120, 0) m/s

Simulation results provide a visually view of the impact brought by topography variations. It is noted that the large QPEs and PEs were eliminated for high-resolution airborne SAR imaging. The focusing approach was a rigorous one in this work, compared with those for the conventional cases which assume that the ground scene is flat.

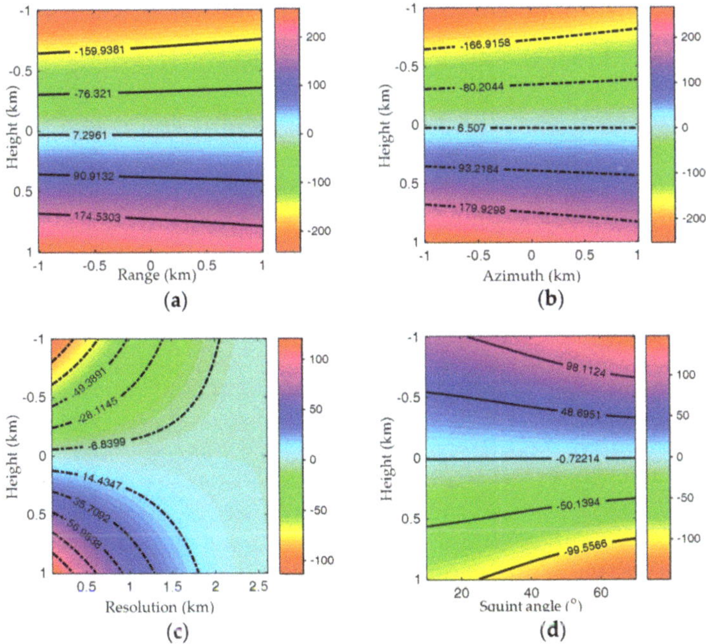

Figure 2. Quadratic phase errors (QPEs) introduced by topography variations in different domains: (**a**) QPEs introduced by topography variations in range/height domain; (**b**) QPEs introduced by topography variations in azimuth/height domain; (**c**) QPEs introduced by topography variations in the resolution/height domain; (**d**) QPEs introduced by topography variations in the squint angle/height domain.

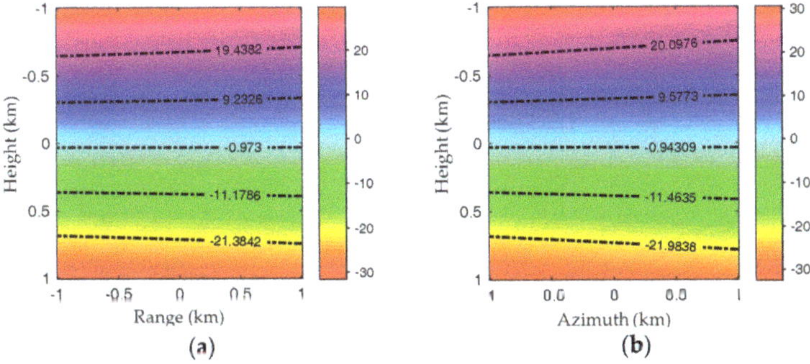

Figure 3. PEs introduced by topography variations in different domains: (**a**) PEs introduced by topography variations in the range/height domain; (**b**) PEs introduced by topography variations in the azimuth/height domain.

For the high-resolution airborne SAR, two main aspects should be considered:

(1) Three-dimensional (3-D) variations in time-domain. Clearly, the general range history in Equation (1) was a vector variable function, which means that one should consider the changes in range, azimuth, and height directions, when designing the focusing approach in the time-domain. An accurate equation indicates the good performance of the proposed focusing method, or it

would lead to a great deterioration in the final imaging result, including the IRW, the peak side-lobe ratio (PSLR), and the integration side-lobe ratio (ISLR).

(2) 3-D variations in the frequency-domain. Because of the complex composition terms in Equation (1), it is difficult to derive the 2-D spectrum, even when using the method of the series reversion (MSR). Moreover, with the complex range history, the focused approach designed in the 2-D frequency-domain would be more complicated than those in the time-domain. Thus, the process directly applied in frequency-domain was not the best choice for an efficient algorithm design for high-resolution airborne SAR in this work.

3. Imaging Algorithm

For the high-resolution airborne SAR, the complex range history in Equation (1) greatly increased the difficulty in imaging, because the 3-D variations that were distributed on the ground scene were difficult to eliminate, especially in the height direction. To solve these problems, a novel focusing algorithm was presented in this section. The proposed EBP algorithm took into account the topography variations, and had the advantages of accuracy and efficiency. The imaging algorithm could be divided into two steps. The first step, i.e., the frequency-domain processing, was used to greatly decrease the virtual aperture length by applying the equivalent time-frequency rotation operations in the frequency-domain. The second step, i.e., the time-domain processing, was to remove the 3-D spatial variations, including the 2-D spatial variations in the horizontal plane, and the spatial variation was brought by topography variations in the vertical plane by performing BP in the time-domain. The specific implementation of these two parts: frequency-domain processing and time-domain processing, was presented as following.

3.1. Frequency-Domain Processing

The defect of the conventional BP algorithm was poor computational efficiency. The reduction of the computational burden would be the key point to improving the imaging efficiency. In this section, a frequency-domain approach was presented to compress the data in an azimuth time-domain for the BP algorithm. The essential of the frequency-domain processing in this work was an equivalent time-frequency rotation operation to change the support area (aperture length) in the azimuth time-domain, which was similar to that of the traditional two-step approach proposed in [47–49]. The difference was that the traditional two-step approach was used to change the equivalent azimuth bandwidth, to avoid spectrum aliasing, whereas the proposed one was to decrease the virtual aperture length, to reduce the computational burden of the BP algorithm. Moreover, they performed in different domains. For instance, the first step of the frequency-domain processing was implemented in the azimuth frequency, while one of the traditional two-step approaches was implemented in the azimuth time-domain. So we first transformed Equation (3) into the range/Doppler domain, i.e., $S_1(t_r, f_\eta)$, where f_η was the azimuth frequency. Similar to that of the two-step approach [47], the reference function was constructed as:

$$H_1(f_\eta) = \exp\left(-j\pi\alpha_0 f_\eta^2\right) \tag{4}$$

where α_0 is the scaling factor and it determines the sampling interval in the time-domain after frequency-domain processing. Convoluting the range/Doppler signal $S_1(t_r, f_\eta)$ of point A with the reference function expressed in Equation (4) yields:

$$
\begin{aligned}
S_2\left(f_\eta'\right) &= S_1\left(t_r, f_\eta'\right) \otimes_{az} H_1\left(f_\eta'\right) \\
&= \underbrace{\exp\left(-j\pi\alpha_0 f_\eta'^2\right)}_{\text{Residual Phase}} \cdot \underbrace{\int S_1\left(t_r, f_\eta\right) \exp\left(-j\pi\alpha_0 f_\eta^2\right)}_{\text{deramping}} \cdot \underbrace{\exp\left(j2\pi\alpha_0 f_\eta f_\eta'\right) df_\eta}_{\text{IFT Kernel}}
\end{aligned} \tag{5}
$$

where $\otimes_{az}$ denotes the azimuth convolution in the frequency-domain. It is clearly that the convolution can be achieved by three operations, i.e., the deramping operation, the inverse Fourier transform

(IFT) with new IFT kernel, and the residual phase compensation, as marked in Equation (5). The frequency-domain processing had the same principle as that of the azimuth preprocessing [49] that was applied in the time-domain, which could be regarded as the 'time-frequency rotation' [48]. According to the new IFT kernel, one can obtain that:

$$\begin{cases} \eta' = \alpha_0 f_\eta \\ \eta = \alpha_0 f_\eta' \end{cases} \tag{6}$$

where η' and f_η' are the new azimuth time and frequency variables, respectively. The residual phase compensation was expressed as:

$$H_2\left(f_\eta'\right) = \exp\left\{-j\pi\alpha_0 f_\eta'^{2}\right\} \tag{7}$$

The convolution in Equation (5) performed in the frequency-domain could be rewritten as a product in the time-domain, i.e.,:

$$S_2\left(t_r, \eta'\right) = S_1\left(t_r, \eta'\right) \cdot H_1\left(\eta'\right) \tag{8}$$

where $H_1(\eta')$ is the time-domain form of $H_1\left(f_\eta'\right)$, and $S_2(t_r, \eta')$ is the form of $S_2\left(f_\eta'\right)$ in the 2-D time domain. To acquire $S_2(t_r, \eta')$, one needs to compensate the second term on the right side of Equation (8), and the compensation function can be constructed as

$$H_3\left(\eta'\right) = \exp\left(-j\pi\eta'^2/\alpha_0\right) \tag{9}$$

Multiplying Equation (9) with Equation (8), one finally obtains the 2-D time-domain expression which has a form that is similar to that of the original echoes, except for different azimuth time variable η'. According to Equation (6), it is noted that the scaling factor α_0 can be selected to decrease support area (virtual aperture length) to reduce the computational burden of the time-domain processing in the next step. Thus, the frequency-domain processing was a key part in this work.

In order to explain the kernel of the frequency-domain processing, the diagram for frequency-domain processing is graphically shown in Figure 4. Four targets are distinguished in the azimuth dimension. The time-frequency diagrams (TFDs) of received raw signal and reference function H_1 are shown in Figure 4a. As shown in Figure 4b, the received signal obtained the new supporting area after the deramping operation, in part of Equation (5). At this step, the echo signal is multiplied with Equation (4) in the azimuth frequency-domain. Then, an equivalent IFT operation with a new IFT kernel in the azimuth from η and f_η to η' and f_η' was implemented, and Figure 4c shows the results of the equivalent IFT operation. At the last step, the compensation was implemented by using Equations (7) and (8) in different domains, and the renovated echo signal was obtained, as shown in Figure 1d. After the operation above, the virtual aperture length of the new echo signal was smaller than the one of original echo signal. This meant that the computational burden of the following algorithm could be decreased, if the data of processed echo signal was selected properly.

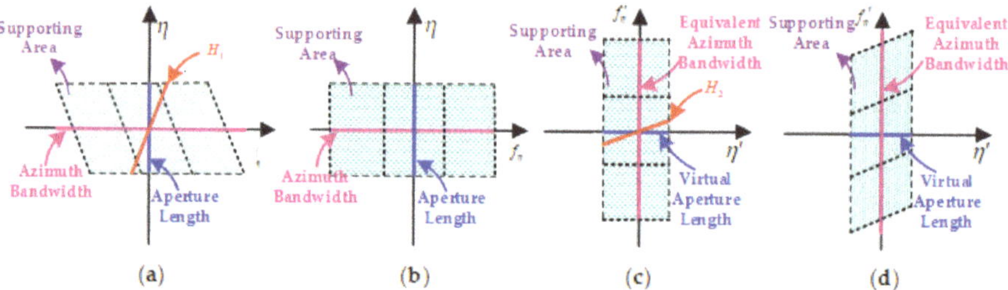

Figure 4. Steps of the frequency-domain processing using the time-frequency diagrams (TFDs). (**a**) Echo signal and reference function H_1 in the range/Doppler domain; (**b**) signal after deramping; (**c**) signal after inverse Fourier transform (IFT) with new IFT kernel and residual phase compensation function H_2; (**d**) signal after the residual phase compensation.

3.2. Time-Domain Processing

The time-domain algorithms are mainly the back projection algorithm (BPA) and its extensions [30–35]. They implement the phase-corrected integration for each imaged point separately, by taking into account the individual propagation delay. So the application of the BPA, and its extensions can easily adapt to airborne SAR with topography variations. However, in terms of computational burden, the BPA is not always the best choice, compared with those of the frequency-domain algorithms. To avoid the heavy computational burden, the combination of the conventional BPA and the frequency-domain processing in Section 3.1 was applied in this work, to improve the efficiency of the imaging algorithm.

After the frequency-domain processing, the echo signal was transformed into the range frequency-domain as:

$$S_3(f_r, \eta') = \varepsilon_0 \omega_r(f_r)\omega_a(\eta') \exp\left(-j\pi\frac{f_r^2}{\gamma}\right) \exp\left(-j\frac{4\pi(f_c + f_r)}{c}r(\eta'; A)\right) \tag{10}$$

Performing the range compression, the compression factor was expressed as:

$$H_4(f_r) = \exp\left(j\pi\frac{f_r^2}{\gamma}\right) \tag{11}$$

Then, multiplying Equation (11) with Equation (10), one can get:

$$\begin{aligned}
S_4(f_r, \eta') &= S_3(f_r, \eta') \cdot H_4(f_r) \\
&= \varepsilon_0 \omega_r(f_r)\omega_a(\eta') \exp\left(-j\frac{4\pi(f_c + f_r)}{c}r(\eta'; A)\right)
\end{aligned} \tag{12}$$

Transforming Equation (12) into 2-D time-domain yields:

$$S_5(t_r, \eta') = \varepsilon_0 G_r G_\eta \mathrm{sinc}\left[\Delta f_r\left(t_r - 2r(\eta'; A)/c\right)\right] \exp\left(-j\frac{4\pi}{\lambda}r(\eta'; A)\right) \tag{13}$$

where Δf_r is the bandwidth of the transmitted signal, G_r and G_η represent the range and the azimuth envelope gain in the frequency-domain respectively, and λ is the wavelength of the transmitted signal.

Based on the new azimuth time η', the BPA was applied in Equation (13) to achieve the focused image. It should be noted that the sampling number in azimuth time-domain can be significantly decreased by selecting properly scaling factor α_0, which can greatly reduce the computational burden with slight deterioration of imaging quality. This was the main reason that we performed the frequency-domain processing first in this work.

Figure 5 shows the illustration of the proposed algorithm. The first step is frequency-domain processing. It was seen that both the virtual aperture length and the new sampling interval were smaller than the original ones, which was decided by the selection of α_0. The length of the original aperture was reduced greatly, and the sampling interval became smaller than the original one. So, some of the positions in the flight path became redundant. In the new aperture, the green spots represented the chosen sampling points, while the red ones represented the redundant sampling points. Thus, a lesser number of positions could be selected for the BPA to reduce the computational burden, as shown in Figure 5. Then, the time-domain processing, i.e., BPA, was applied in the 2-D time-domain. It was important to note that the DEM data should be known to achieve the accurate range history.

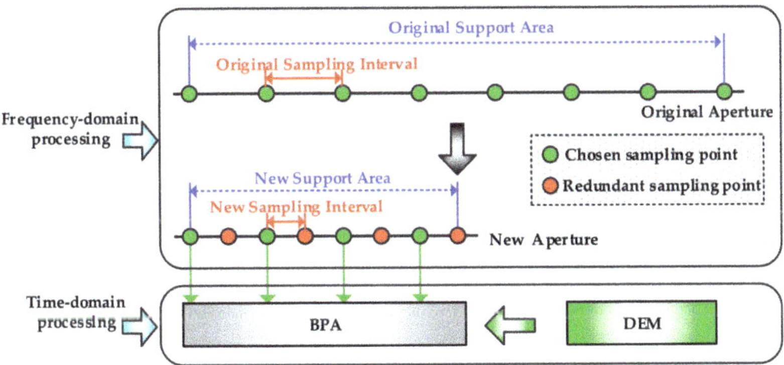

Figure 5. Illustration of proposed algorithm.

3.3. Flowchart of the EBP Algorithm

Based on the aforementioned discussion, Figure 6 presents the flowchart of the EBP algorithm. H_1, H_2, H_3, and H_4 were respectively expressed as Equations (4), (7), (9) and (11). Clearly, the proposed EBP algorithm included two parts: frequency-domain processing was used to reduce the length of the virtual aperture of the received signal and the computational burden of the following algorithm; time-domain processing solves the problems that are caused by topography variations, with the combination of the selection of the sampling points, the DEM data, and the back-projection algorithm. The processing procedure of the proposed algorithm was practically denoted as follows:

1. Azimuth Fourier transform (FT). Apply an azimuth FT on the raw data;
2. Deramping processing; the first step of the frequency-domain processing. Modulate the raw data with Equation (4);
3. Equivalent Azimuth IFT; the second step of the frequency-domain processing. Apply the transformation in the data from the η and f_η coordinate, to the η' and f'_η coordinate;
4. Residual phase compensation; the final step of the frequency-domain processing. Compensate the data with Equation (7) in the new azimuth frequency-domain;
5. Azimuth IFT. Transform the data from the f'_η domain to η' domain;
6. Second phase compensation. Compensate the data with Equation (9) in the new azimuth time-domain;
7. Range FT. Apply a range of FT on the renovated data. Then, the range compression will be implemented in the range frequency-domain;
8. Range compression. Compress the data in the range frequency-domain with Equation (11);
9. Range IFT. Transform the data from the t_r domain to f_r domain;

10. Selection of samples. The important sampling points are selected appropriately, and the premise is that the echo signal will not alias the azimuth frequency-domain. In this way, the computational burden of the BP algorithm will be reduced;

11. BP processing. Calculate the accurate slant range from each azimuth position to the targets, with the combination of the airborne acquisition scenarios and the DEM data. The other substeps in this step are similar to those of the classical BP processing.

For the purpose of simplicity and clarity, the work mode of the high-resolution airborne SAR system is designed as a spotlight SAR in the previous discussion of the proposed algorithm. Meanwhile, it is assumed that the airborne SAR flies along the linear trajectory. Actually, the proposed EBP algorithm can be also applied in other work modes, such as the stripmap SAR, terrain observation by progressive scans (TOPS) SAR, or the sliding spotlight SAR. These modes are distinguished, based on the variation of the antenna beam direction, and they are all called beam steering SAR for simplicity [50]. It is well known that the time-domain back-projection algorithm can be considered as a linear transformation from radar echo data to the SAR scene; thus, the proposed algorithm based on the BP algorithm could be applied in SAR, with the above-mentioned modes. Furthermore, the EBP algorithm was also applicable to the other SAR system (including the airborne SAR and spaceborne SAR), flying along the curve trajectory.

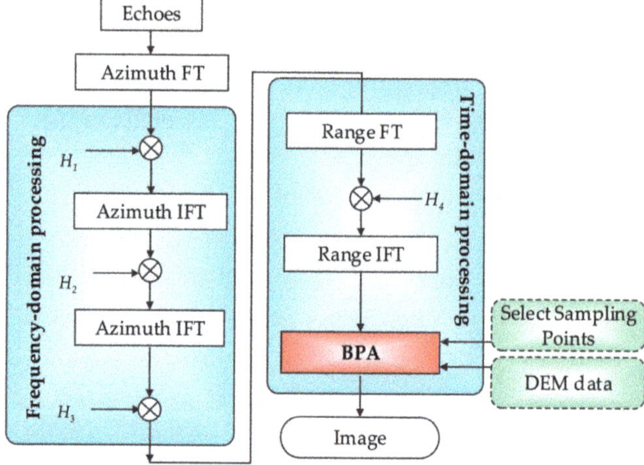

Figure 6. The flowchart of the EBP algorithm. In this flowchart, frequency-domain processing and time-domain processing are included.

4. Simulation Results

In this section, to prove the efficiency of EBP algorithm, spot matrix simulation results and SAR scene simulation results are respectively presented.

4.1. Spot Matrix Simulation Results

In this subsection, the performance of the proposed EBP algorithm is evaluated in the cases without, and with consideration of the motion errors. The parameters for the simulation are listed in Table 1.

Figure 7 is the distribution of the targets in the illuminated spot matrix. The scene size (range × azimuth) is about 1.2 km × 1.2 km. The targets in the different location of the scene was at different altitudes.

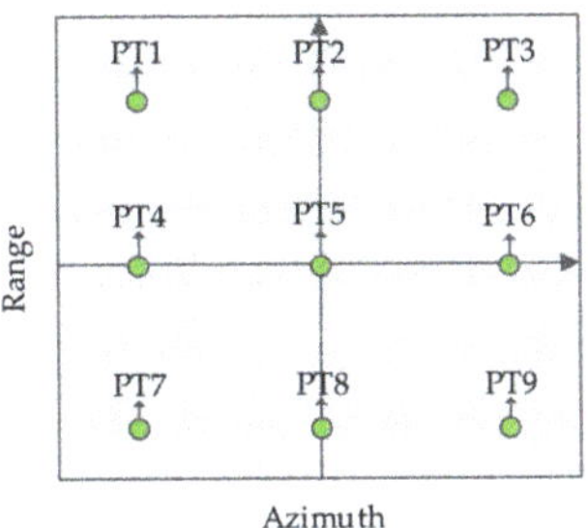

Figure 7. Distribution of targets in the illuminated spot matrix which contained nine-point targets.

Case I

In this experiment, simulation results without considering the topography variations were used for comparisons. The airborne SAR worked in the spotlight mode. The motion errors were not added in this case. The heights of targets PT1 and PT9 were respectively set as -120.0 and 180.0 m, with respect to the reference point PT5, the squint angle was $30°$, and the azimuth resolution was 0.681 m. The traditional chirp scaling algorithm (CSA) (e.g., [15]) was used for comparison.

Figure 8 shows the comparative results by the proposed algorithm and CSA. From Figure 8, we see that the topography variations in the vertical direction were greatly decreased, and the targets with different range and azimuth positions by the proposed approach were visually well-focused. However, for the CSA without considering the topography variations in the vertical direction, although the reference target in the scene center could be well-focused, the targets on the edges remained highly defocused for the reason, that the errors introduced by the target heights cannot be ignored in the high-resolution case.

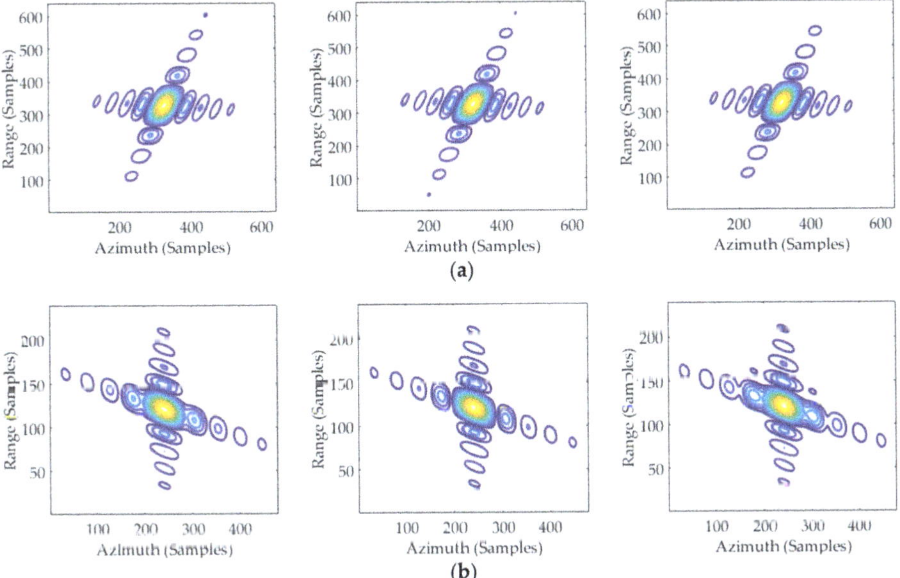

Figure 8. Comparative results by proposed algorithm and CSA. (Left to right) Targets PT1, PT5, and PT9, respectively. (**a**) Contour plot of targets (PT1, PT5, and PT9) processed by the EBP algorithm; (**b**) Contour plot of targets (PT1, PT5, and PT9) processed by CSA.

To acquire the further comparison of the focusing performance, a quantitative analysis with IRW, PSLR, and ISLR was used as criteria. The corresponding imaging performances of the targets PT1, PT5, and PT9 are calculated, and the results of proposed algorithm, and the CSA were listed in Table 2. Both the contour results and imaging quality parameters of the proposed algorithm approach the theoretical values of IRW (0.681 m), PSLR (−13.26 dB), and ISLR (−9.8 dB), which indicated that the proposed algorithm could be well-applied to the high-resolution airborne SAR with topography variations. On the other hand, the corresponding parameters that were obtained by the CSA were relatively worse, especially for the targets in the edge of the scene, which were much less than the theoretical values. It means that the CSA was inferior to the proposed algorithm.

Table 2. Imaging quality parameters in Simulation Case I.

Method	Target	IRW (m)	PSLR (dB)	ISLR (dB)
Proposed algorithm	PT1	0.687	−13.22	−10.03
	PT5	0.681	−13.27	−10.05
	PT9	0.679	−13.24	−10.02
CSA	PT1	0.795	−7.23	−6.65
	PT5	0.682	−13.25	−10.06
	PT9	0.813	−7.14	−6.37

Case II

In this experiment, the heights of targets PT1 and PT9 were respectively 452.5 and −231.1 m, with respect to target PT5, which was set as the reference point in Figure 7. The squint angle was 60°, and azimuth resolution was 0.472 m. The PTA-MoComp approach (PMA) [25] was performed for comparisons. Figure 9a is the motion errors extracted from the airborne inertial navigation system. Figure 9b shows the focused result of the simulation spot matrix. It is well-noted that the positions of targets PT1 and PT9 were well-located without distortions in the image.

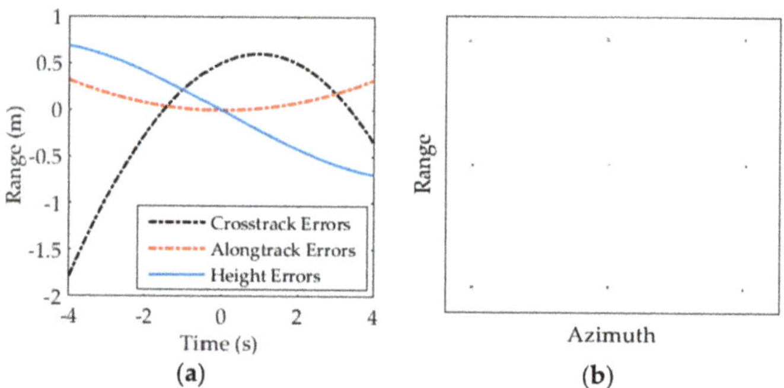

Figure 9. Motion errors and simulation results: (**a**) motion errors extracted from the airborne inertial navigation system; (**b**) focused result of the simulation spot matrix by the proposed algorithm.

Figure 10 shows the comparative results of the azimuth profiles focused by PMA and the proposed algorithm. It was well-noted that the targets were visually well-focused, with a relatively clear separation of the main lobes and the first and subsequential sidelobe, by the proposed algorithm. Since there was no extra windowing operation, the theoretical values of PSLR were about −13 dB. Clearly, the proposed algorithm had a good performance (red dashed lines in figures). For the traditional method of PMA, the reference target could be well focused; however, the edge targets had great degradations because the post azimuth-matched filter was sensitive to azimuth-dependent motion errors. The imaging quality parameters were listed in Table 3. Imaging quality parameters of the proposed algorithm were obviously close to the theoretical values of IRW (0.472 m), PSLR (−13.26 dB), and ISLR (−9.8 dB), while the ones of the PMA were relatively worse. Our algorithm was noted to be generally superior to the PMA. Consequently, with the incorporation of the frequency-domain processing into the time-domain processing, promising results were obtained for high-resolution airborne SAR.

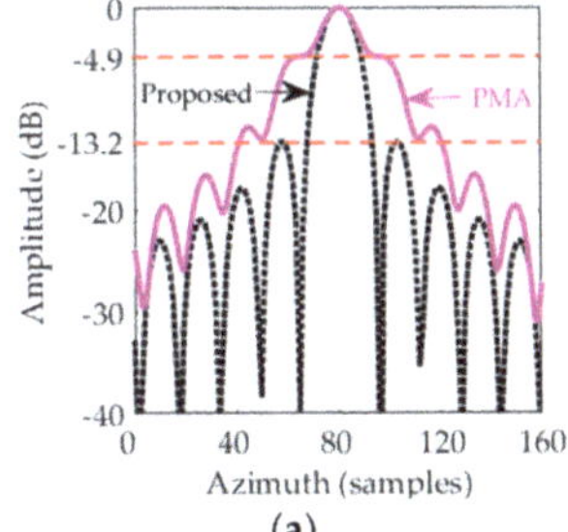
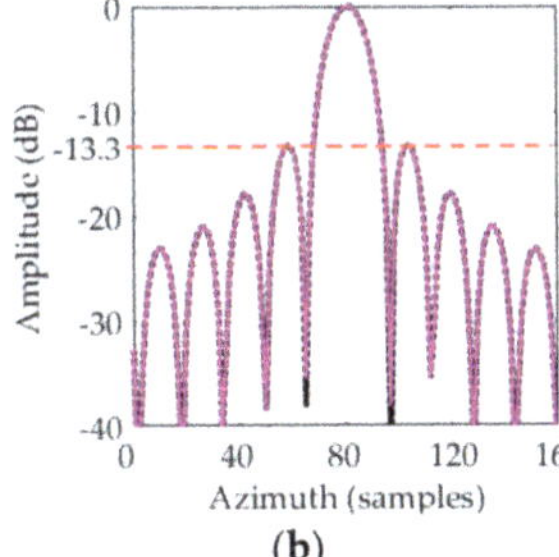
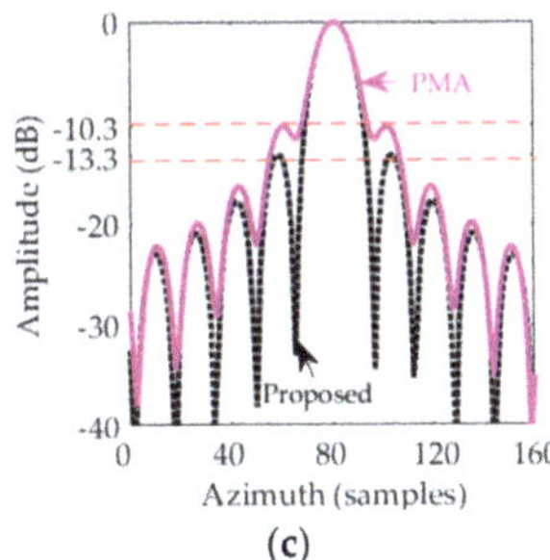

Figure 10. Simulation results: (**a**) comparative results of azimuth profiles of PT1 by PMA and the proposed algorithm; (**b**) comparative results of azimuth profiles of PT5 by PMA and the proposed algorithm; (**c**) comparative results of azimuth profiles of PT9 by PMA and the proposed algorithm.

Table 3. Imaging quality parameters in Simulation Case II.

Method	Target	IRW (m)	PSLR (dB)	ISLR (dB)
	PT1	0.474	−13.27	−10.07
Proposed algorithm	PT5	0.472	−13.31	−10.04
	PT9	0.471	−13.30	−10.01
	PT1	0.613	−4.93	−4.64
PMA	PT5	0.423	−13.32	−10.06
	PT9	0.546	−10.35	−7.58

4.2. Scene Simulation Results

Corresponding to the experiment of the spot matrix simulation, the scene simulation results were presented in this subsection. The echo data of the scene was obtained by the time-domain simulation method, with motion errors as shown in Figure 9a. The imaging results of the echo data validated the efficiency of the proposed EBP algorithm. Here, it was assumed that the data of an X-band SAR with the squinted spotlight mode was used to demonstrate the performance of the proposed algorithm. The center squint angle was about 60°, and the velocity was about 120 m/s. The whole scene was about 1.2 km in the cross-range, and 1.5 km in the range. Figure 11 showed the focusing results, which were processed by traditional BPA, the proposed algorithm, and PMA, respectively. Particularly, the topography variations were added into the dotted line rectangular region in Figure 11. The difference between the highest and the lowest points in the topography variations was 310.36 m.

The entire scene was well focused, with the proposed algorithm in Figure 11b, including the edge regions. In order to obtain a clear and detailed comparison, the local enlargements of the solid line rectangular regions of each image were displayed respectively in the dashed line rectangular regions. It was evident that the imaging quality in Figure 11b was as good as the one in Figure 11a, and it was better than the one in Figure 11c. Therefore, the proposed algorithm could handle the effects of the topography variations well, for high-resolution airborne SAR.

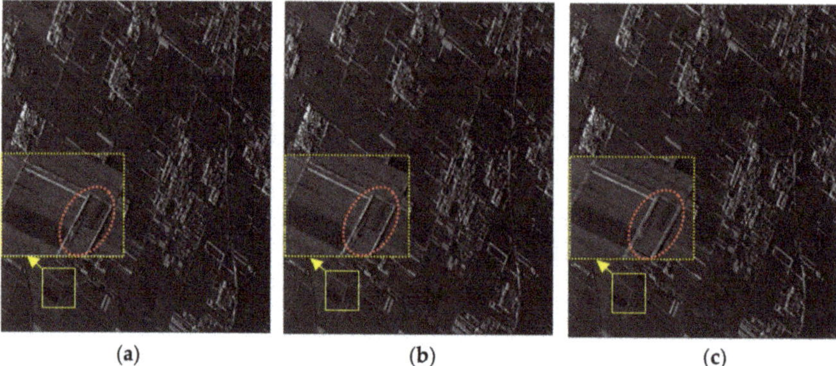

(a)   (b)   (c)

Figure 11. Scene simulation data results processed by different algorithms. (**a**) Scene simulation data results processed by traditional BPA. (**b**) Scene simulation data results processed by the EBP algorithm. (**c**) Scene simulation data results processed by the reference algorithm.

5. Discussion

5.1. Parameter Selection

In the Section 3.1, the echo signal may have aliasing after the frequency-domain processing, if the scaling factor α_0 was selected improperly. To weaken the signal aliasing, the zero padding operation was taken in some literatures. However, the zero padding would increase the computation burden of the whole algorithm, which was the opposite of the purpose of the proposed algorithm. Therefore, proper selection for the parameter α_0 without zero padding operation was necessary for frequency-domain processing. As has been discussed in [51], the two-step processing could be regarded as a special case of two Fractional Fourier transform (FrFT) operations. Similarly, the frequency-domain processing could also be implemented by two FrFT operations with parameters, χ and κ. We obtained the relation between the parameters as:

$$\chi + \kappa = \frac{\pi}{2} \tag{14}$$

$$\cot \chi = -\pi \alpha_0 \tag{15}$$

After frequency-domain processing, the whole bandwidth of the echo signal should be less than the pulse repetition frequency (PRF), in the case of the phenomenon of frequency aliasing. It means that:

$$T_a |\cot \eta_0 + \tan \chi| + 2\pi B_s \leq 2\pi PRF \tag{16}$$

$$T_a |\cot \eta_0| + 2\pi B_s \leq \frac{m'}{m} T_a |\tan \chi| \tag{17}$$

where η_0 is the original slant angle of a point target in the time-frequency distribution, B_s is the instantaneous bandwidth of the echo signal in the receiving terminal, m and m' denote the numbers of sampling points in the original azimuth time-domain and the new azimuth time-domain, respectively.

The selection of parameter α_0 is usually obtained through the Equations (14)–(17). In general, the parameter should be selected, with the guarantee that the supporting area of the rotated signal should be in its available area.

5.2. Computational Burden

It is well-known that the BPA can precisely manage the problems caused by the curved path model. However, the BPA has a much heavier computational burden than that of the frequency-domain algorithms. In essence, the proposed algorithm in this paper was an improved BP algorithm, so it was necessary to make discussion about the computational burden of the proposed algorithm here. For simplification and without loss of generality, the computational burden of the traditional CS was compared as the standard.

The frequency-domain processing in this work was essentially a frequency-domain algorithm. In the frequency-domain processing, three azimuth fast Fourier transforms (FFTs), and three multiplications are utilized. Therefore, the total computational burden of the frequency-domain processing could be expressed as:

$$O_F = 1.5NM \log_2 M + 3NM \tag{18}$$

where M is the number of the azimuth samples, and N is the number of the range samples. Moreover, the computational burden of time-domain processing presented in Section 3.2 is discussed. Considering the focusing algorithm, i.e., two range FFTs, one multiplication, and BPA, the computational burden of the time-domain processing can be derived as:

$$O_T = MN + NM \log_2 N + \mu_{\text{int}} M^2 N \tag{19}$$

where μ_{int} denotes the proportional factor of the computational burden of the post-processing, which is determined by the rotation factor. Thus, with Equations (18) and (19), the total computational burden of the EBP algorithm was expressed as:

$$O_s = 1.5NM \log_2 M + 4NM + NM \log_2 N + \mu_{\text{int}} M^2 N \tag{20}$$

In comparison, of the BPA (e.g., [52]), for each pixel in the final image, a signal vector with a length M was extracted from the range-compressed data, multiplied with a phase function, and summed; thus, the total computational burden could be expressed as:

$$O_B = M^2 N \tag{21}$$

In a standard CSA, two range FFTs, one azimuth FFT, and three multiplications are used. Therefore, the computational burden of the standard CSA can be expressed as:

$$O_C = MN \log_2 N + NM \log_2 M + 3NM \tag{22}$$

The results of the comparison mentioned above with specific parameters, are shown in Table 4, based on Equations (20)–(22). As listed in Table 4, ratios at five sample sizes were computed to quantify the comparison. Clearly, the computational burden of the BPA was much larger than that of the proposed algorithm, which meant that our method could greatly improve the operation speed. Even the computational burden of the proposed algorithm was slightly larger than that of the CSA, but there was wide application in the high-resolution airborne SAR, with topography variations.

Table 4. Computational load analysis using CSA as references.

Data Size in Azimuth ($\times 10^3$)	1	2	4	8	16
BPA/CSA	150.30	290.11	560.73	1084.92	2101.48
Proposed/CSA	12.07	22.15	41.64	79.39	152.59

6. Conclusions

The requirement for the resolution for remote sensing has become higher and higher. When the resolution is at the decimeter level, topography variations were ignored in conventional airborne SAR would cause serious phase errors. For the high-resolution airborne SARs with topography variations, 3-D variations in time or frequency-domains were difficult to eliminate with the traditional algorithms. Especially for the topography variations, the expression of the received echo data in the frequency-domain was difficult to derive. In this paper, based on the precise physical model of airborne SAR considering the topography variations, phase errors caused by topography variations are analyzed in different domains first.

The proposed EBP algorithm includes two main parts: frequency-domain processing and time-domain processing. The frequency-domain processing is essentially an operation of time-frequency rotation to reduce the virtual aperture length greatly, by selecting a suitable rotation factor. The computational burden can be decreased by removing the redundant position in the flight track. Then the accurate range history is obtained with DEM data in the time-domain processing. The BP algorithm can be applied directly in the 2-D time-domain with an accurate range history. In the simulation, the spot matrix simulation results and the scene simulation data validated the performance of the EBP algorithm. Finally, the computational burden of EBP algorithm was compared with those of other algorithms. It was obvious that our algorithm can greatly improve the operation speed compared with the traditional BPA.

Author Contributions: C.L. and S.T. conceived the main idea; S.T., L.Z., and P.G. conceived and designed the experiments; C.L. and S.T. analyzed the data and wrote the paper.

Acknowledgments: This work was supported in part by the National Natural Science Foundation of China under Grants 61601343, 61701393, 61731023, and 61671361, in part by the China Postdoctoral Science Foundation Funded Project under Grant 2016M600768, in part by the National Natural Science Foundation of the Shaanxi Province under grant 2018JM6054, and in part by the National Defense Foundation of China.

Conflicts of Interest: The authors declare no conflict of interest.

References

1. Cumming, I.G.; Wong, F.H. *Digital Processing of Synthetic Aperture Radar Data: Algorithm and Implementation*; Artech House: Norwood, MA, USA, 2005.
2. Sun, H.; Shimada, M.; Xu, F.; Xu, F. Recent Advances in Synthetic Aperture Radar Remote Sensing—Systems, Data Processing, and Applications. *IEEE Geosci. Remote Sens. Lett.* **2017**, *14*, 2013–2016. [CrossRef]
3. Carrara, W.G.; Majewski, R.M.; Goodman, R.S. *Spotlight Synthetic Aperture Radar: Signal Processing Algorithms*; Artech House: Norwood, MA, USA, 1995.
4. Dudczyk, J.; Kawalec, A. Optimizing the Minimum Cost Flow Algorithm for the Phase Unwrapping Process in SAR Radar. *Bull. Pol. Acad. Sci. Tech. Sci.* **2014**, *62*, 511–516. [CrossRef]
5. Matuszewski, J. The Analysis of Modern Radar Signals Parameters in Electronic Intelligence System. In Proceedings of the 13th International Conference on Modern Problems of Radio Engineering, Telecommunications and Computer Science (TCSET), Lviv, Ukraine, 23–26 February 2016; pp. 298–302.
6. Matuszewski, J.; Paradowski, L. The Knowledge Based Approach for Emitter Identification. In Proceedings of the 12th International Conference on Microwaves and Radar, Krakow, Poland, 20–22 May 1998; pp. 810–814.
7. Soumekh, M. *Synthetic Aperture Radar Signal Processing with MATLAB Algorithms*; Wiley-Interscience: Hoboken, NJ, USA, 1999.
8. Zhang, T.; Ding, Z.; Tian, W.; Zeng, T. A 2-D nonlinear chirp scaling algorithm for high squint GEO SAR imaging based on optimal azimuth polynomial compensation. *IEEE J. Sel. Top. Appl. Earth Observat. Remote Sens.* **2017**, *10*, 5724–5735. [CrossRef]
9. Tang, S.; Zhang, L.; Guo, P.; Liu, G.; Zhang, Y.; Li, Q.; Gu, Y.; Lin, C. Processing of monostatic SAR data with general configurations. *IEEE Trans. Geosci. Remote Sens.* **2015**, *53*, 6529–6546.
10. Sun, G.C.; Wu, Y.; Wang, Y.; Xing, M.D.; Bao, Z. Full-aperture focusing of very high resolution spaceborne-squinted sliding spotlight SAR data. *IEEE Trans. Geosci. Remote Sens.* **2017**, *55*, 3309–3321. [CrossRef]
11. Hobbs, S.; Mitchell, C.; Forte, B.; Holley, R.; Snapir, B.; Whittaker, P. System design for geosynchronous synthetic aperture radar missions. *IEEE Trans. Geosci. Remote Sens.* **2014**, *52*, 7750–7763. [CrossRef]
12. Wang, C.; Chen, L.; Liu, L. A New Analytical Model to Study the Ionospheric Effects on VHF/UHF Wideband SAR Imaging. *IEEE Trans. Geosci. Remote Sens.* **2017**, *55*, 4545–4557. [CrossRef]

13. Moreira, A.; Krieger, G.; Hajnsek, I.; Papathanassiou, K.P.; Younis, M.; Lopez-Dekker, P.; Huber, S.; Villano, M.; Pardini, M.; Eineder, M.; et al. Tandem-L: A highly innovative bistatic SAR mission for global observation of dynamic processes on the Earth's surface. *IEEE Geosci. Remote Sens. Mag.* **2015**, *3*, 8–23. [CrossRef]

14. Fornaro, G.; Franceschetti, G.; Perna, S. Motion Compensation Errors: Effects on the accuracy of airborne SAR Images. *IEEE Trans. Aerosp. Electron. Syst.* **2005**, *41*, 1338–1352. [CrossRef]

15. Raney, R.K.; Runge, H.; Bamler, R.; Cumming, I.G.; Wong, F.H. Precision SAR processing using chirp scaling. *IEEE Trans. Geosci. Remote Sens.* **1994**, *32*, 786–799. [CrossRef]

16. Cafforio, C.; Prati, C.; Rocca, E.C. SAR focusing using seismic migration techniques. *IEEE Trans. Aerosp. Electron. Syst.* **1991**, *27*, 194–207. [CrossRef]

17. Sun, G.C.; Jiang, X.; Xing, M.D.; Qiao, Z.J.; Wu, Y.; Bao, Z. Focus improvement of highly squinted data based on azimuth nonlinear scaling. *IEEE Trans. Geosci. Remote Sens.* **2011**, *49*, 2308–2322. [CrossRef]

18. Wong, F.W.; Yeo, T.S. New applications of nonlinear chirp scaling in SAR data processing. *IEEE Trans. Geosci. Remote Sens.* **2001**, *39*, 946–953. [CrossRef]

19. Zeng, T.; Hu, C.; Wu, L.; Liu, L.; Tian, W.; Zhu, M.; Long, T. Extended NLCS algorithm of BiSAR systems with a squinted transmitter and a fixed receiver: Theory and experimental confirmation. *IEEE Trans. Geosci. Remote Sens.* **2013**, *51*, 5019–5030. [CrossRef]

20. Tang, S.; Zhang, L.; So, H.C. Focusing high-resolution highly-squinted airborne SAR data with maneuvers. *Remote Sens.* **2018**, *10*, 862. [CrossRef]

21. Qiu, X.; Hu, D.; Ding, C. An improved NLCS algorithm with capability analysis for one-stationary BiSAR. *IEEE Trans. Geosci. Remote Sens.* **2008**, *46*, 3179–3186. [CrossRef]

22. Wang, Y.; Li, J.; Yang, J. Wide Nonlinear chirp scaling algorithm for spaceborne stripmap range sweep SAR imaging. *IEEE Trans. Geosci. Remote Sens.* **2017**, *55*, 6922–6936. [CrossRef]

23. Sun, G.C.; Xing, M.D.; Wu, Y.; Wang, Y.; Yang, J.; Bao, Z. A 2-D space-variant chirp scaling algorithm based on the RCM equalization and subband synthesis to process geosynchronous SAR data. *IEEE Trans. Geosci. Remote Sens.* **2014**, *52*, 4868–4880.

24. Li, D.; Lin, H.; Liu, H.; Liao, G.; Tan, X. Subaperture approach based on azimuth-dependent range cell migration correction and azimuth focusing parameter equalization for maneuvering high-squintmode SAR. *IEEE Trans. Geosci. Remote Sens.* **2017**, *53*, 6718–6734.

25. Macedo, K.A.C.; Scheiber, R. Precise topography- and aperture-dependent motion compensation for airborne SAR. *IEEE Trans. Geosci. Remote Sens.* **2005**, *2*, 172–176.

26. Prats, P.; Reigber, A.; Mallorqui, J.J. Topography-dependent motion compensation for repeat-pass interferometric SAR systems. *IEEE Trans. Geosci. Remote Sens.* **2005**, *2*, 206–210. [CrossRef]

27. Yang, Z.M.; Sun, G.C.; Xing, M.D. A new fast back-projection algorithm using polar format algorithm. In Proceedings of the Asia-Pacific Conference on Synthetic Aperture Radar (APSAR), Tsukuba, Japan, 23–27 September 2013; pp. 373–376.

28. Rodriguez-Cassola, M.; Prats, P.; Krieger, G.; Moreira, A. Efficient time-domain image formation with precise topography accommodation for general bistatic SAR configurations. *IEEE Trans. Aerosp. Electron. Syst.* **2011**, *47*, 2949–2966. [CrossRef]

29. Gazdag, J.; Sguazzero, P. Migration of Seismic Data. *Proc. IEEE.* **1984**, *72*, 1302–1315. [CrossRef]

30. Seger, O.; Herberthson, M.; Hellsten, H. Real-time SAR processing of low frequency ultra wideband radar data. In Proceedings of the EUSAR, Friedrichshafen, Germany, 25–27 May 1998.

31. Ferretti, A.; Prati, C.; Rocca, F. Multibaseline InSAR DEM reconstruction: The wavelet approach. *IEEE Trans. Geosci. Remote Sens.* **1999**, *37*, 705–715. [CrossRef]

32. Ferraiuolo, G.; Meglio, F.; Pascazio, V.; Schirinzi, G. DEM reconstruction accuracy in multichannel SAR interferometry. *IEEE Trans. Geosci. Remote Sens.* **2009**, *47*, 191–201. [CrossRef]

33. Shao, Y.F.; Wang, R.; Deng, Y.K.; Liu, Y.; Chen, R.; Liu, G.; Balz, T.; Loffeld, O. Digital elevation model reconstruction in multichannel spaceborne/stationary SAR interferometry. *IEEE Geosci. Remote Sens. Lett.* **2014**, *11*, 2080–2084. [CrossRef]

34. Nilsson, S. Application of Fast Backprojection Techniques for Some Inverse Problems in Integral Geometry. Ph.D. Thesis 499, Linkoping University, Linkoping, Sweden, 1997.

35. Ulander, L.; Hellsten, H.; Stenström, G. Synthetic-aperture radar processing using fast factorized back-projection. *IEEE Trans. Aerosp. Electron. Syst.* **2003**, *39*, 760–776. [CrossRef]

36. Cao, N.; Lee, H.; Zaugg, E.; Shrestha, R.; Carter, W.; Glennie, C.; Wang, G.; Lu, Z.; Fernandez-Diaz, J.C. Airborne DInSAR results using time-domain backprojection algorithm: A case study over the slumgullion landslide in Colorado with validation using spaceborne SAR, airborne LiDAR, and ground-based observations. *IEEE J. Sel. Topics Appl. Earth Observ. Remote Sens.* **2017**, *108*, 4987–5000. [CrossRef]
37. Heng, Z.; Jiangwen, T.; Robert, W.; Yunkai, D.; Wei, W.; Ning, L. An accelerated backprojection algorithm for monostatic and bistatic SAR processing. *Remote Sens.* **2018**, *10*, 140. [CrossRef]
38. Tang, J.; Deng, Y.; Wang, R.; Zhao, S.; Li, N.; Wang, W. A weighted backprojection algorithm for azimuth multichannel SAR imaging. *IEEE Geosci. Remote Sens. Lett.* **2016**, *13*, 1265–1269. [CrossRef]
39. Yang, L.; Zhou, S.; Zhao, L.; Xing, M. Coherent Auto-Calibration of APE and NsRCM under Fast Back-Projection Image Formation for Airborne SAR Imaging in Highly-Squint Angle. *Remote Sens.* **2018**, *10*, 321. [CrossRef]
40. Vu, V.T.; Sjogren, T.K.; Pettersson, M.I. Fast time-domain algorithms for UWB bistatic SAR processing. *IEEE Trans. Aerosp. Electron. Syst.* **2013**, *49*, 1982–1994. [CrossRef]
41. Walterscheid, I.; Espeter, T.; Brenner, A.R.; Klare, J.; Ender, J.H.G.; Nies, H.; Wang, R.; Loffeld, O. Bistatic SAR experiments with PAMIR and TerraSAR-x; setup, processing, and image results. *IEEE Trans. Geosci. Remote Sens.* **2010**, *48*, 3268–3279. [CrossRef]
42. Shao, Y.; Wang, R.; Deng, Y.; Liu, Y.; Chen, R.; Liu, G.; Loffeld, O. Fast backprojection algorithm for bistatic SAR imaging. *IEEE Geosci. Remote Sens. Lett.* **2013**, *10*, 1080–1084. [CrossRef]
43. Duque, S.; Lopez-Dekker, P.; Mallorqui, J.J. Single-pass bistatic SAR interferometry using fixed-receiver configurations: Theory and experimental validation. *IEEE Trans. Geosci. Remote Sens.* **2010**, *48*, 2740–2749. [CrossRef]
44. Peng, X.; Wang, Y.; Hong, W.; Wu, Y. Autonomous Navigation Airborne Forward-Looking SAR High Precision Imaging with Combination of Pseudo-Polar Formatting and Overlapped Sub-Aperture Algorithm. *Remote Sens.* **2013**, *11*, 6063–6078. [CrossRef]
45. Xu, W.; Deng, Y.; Huang, P.; Wang, R. Full-aperture SAR data focusing in the spaceborne squinted sliding-spotlight mode. *IEEE Trans. Geosci. Remote Sens.* **2014**, *52*, 4596–4607. [CrossRef]
46. Tang, S.; Zhang, L.; Guo, P.; Liu, G.; Sun, G.C. Acceleration model analyses and imaging algorithm for highly squinted airborne spotlight mode SAR with maneuvers. *IEEE J. Sel. Topics Appl. Earth Observ. Remote Sens.* **2015**, *8*, 1120–1131. [CrossRef]
47. Lanari, R.; Tesauro, M.; Sansosti, E.; Fornaro, G. Spotlight SAR data focusing based on a two-step processing approach. *IEEE Trans. Geosci. Remote Sens.* **2001**, *39*, 1993–2004. [CrossRef]
48. Yang, M.D.; Zhu, D.Y.; Song, W. Comparison of two-step and one-step motion compensation algorithms for airborne synthetic aperture radar. *Electron. Lett.* **2015**, *51*, 1108–1110. [CrossRef]
49. Lanari, R.; Zoffoli, S.; Sansosti, E.; Fornaro, G.; Serafino, F. New approach for hybrid strip-map/spotlight SAR data focusing. *Proc. Inst. Elect. Eng.-Radar Sonar Navig.* **2001**, *148*, 363–372. [CrossRef]
50. Sun, G.C.; Jiang, X.; Xing, M.D.; Xia, X.G.; Wu, Y.R.; Bao, Z. Beam steering SAR data processing by a generalized PFA. *IEEE Trans. Geosci. Remote Sens.* **2013**, *51*, 4366–4377. [CrossRef]
51. Sun, G.C.; Xing, M.D.; Xia, X.G.; Yang, J.; Wu, Y.R.; Bao, Z. A unified focusing algorithm for several modes of SAR based on FrFT. *IEEE Trans. Geosci. Remote Sens.* **2013**, *51*, 3139–3155. [CrossRef]
52. Xie, H.; Shi, S.; An, D.; Wang, G.; Wang, G.; Xiao, H.; Huang, X.; Zhou, Z.; Xie, C.; Wang, F.; et al. Fast factorized backprojection algorithm for one-stationary bistatic spotlight circular SAR image formation. *IEEE J. Sel. Top. Appl. Earth Observ. Remote Sens.* **2017**, *10*, 1494–1510. [CrossRef]

Article

Coherence-Factor-Based Rough Surface Clutter Suppression for Forward-Looking GPR Imaging

Davide Comite [1], **Fauzia Ahmad** [2,*], **Traian Dogaru** [3] **and Moeness Amin** [4]

[1] Department of Information Engineering, Electronics and Telecommunications, Sapienza University of Rome, 00185 Rome, Italy; davide.comite@uniroma1.it
[2] Department of Electrical and Computer Engineering, Temple University, Philadelphia, PA 19122, USA
[3] U.S. Army Research Lab, Adelphi, MD 20783, USA; traian.v.dogaru.civ@mail.mil
[4] Center for Advanced Communications, Villanova University, Villanova, PA 19085, USA; moeness.amin@villanova.edu
* Correspondence: fauzia.ahmad@temple.edu

Received: 1 February 2020; Accepted: 3 March 2020; Published: 6 March 2020

Abstract: We present an enhanced imaging procedure for suppression of the rough surface clutter arising in forward-looking ground-penetrating radar (FL-GPR) applications. The procedure is based on a matched filtering formulation of microwave tomographic imaging, and employs coherence factor (CF) for clutter suppression. After tomographic reconstruction, the CF is first applied to generate a "coherence map" of the region in front of the FL-GPR system illuminated by the transmitting antennas. A pixel-by-pixel multiplication of the tomographic image with the coherence map is then performed to generate the clutter-suppressed image. The effectiveness of the CF approach is demonstrated both qualitatively and quantitatively using electromagnetic modeled data of metallic and plastic shallow-buried targets.

Keywords: forward-looking GPR; surface clutter; near-field; antenna arrays; microwave imaging; coherence factor

---

1. Introduction

Microwave imaging has undergone significant advances in the last two decades, owing to its increased adoption and broad application in a variety of disciplines, including applied geophysics, planetary exploration, and emerging radar technologies [1–7]. Forward-looking ground penetrating radar (FL-GPR) is one such technology that employs microwave imaging for detection of targets buried at shallow depths in the ground. Unlike its ground-coupled or near-ground down-looking ground penetrating radar (DL-GPR) counterparts, FL-GPR provides standoff sensing capability, which allows fast scanning of large areas for real-time target detection. This capability, however, comes at the expense of energy backscattered from the illuminated targets and limited image spatial resolution [8–15]. Further, the rough ground surface generates clutter that tends to obscure the buried targets, rendering target detection difficult and challenging [12–14].

Rough surface clutter suppression for array-based FL-GPR imaging was addressed in [11,14,16–24]. In [16], an ambiguity function based detector was proposed which exploits time-frequency characterization of target and clutter scattering for performance enhancement. Frequency subband processing was exploited in [11] to obtain the best contrast between target and clutter signals, whereas recursive side-lobe minimization algorithm for reconstructing FL-GPR images with reduced clutter was proposed in [17]. Coherent integration of measurements corresponding to multiple radar platform positions was demonstrated in [18] for rough surface clutter suppression, whereas a nonlinear combining approach that exploits a similarity measure was developed in [19] to adaptively mitigate imaging artifacts. In [20], localized clutter outside of the region of interest was suppressed prior to sparse reconstruction. A real-time three-dimensional (3-D) model of the rough surface scattering was proposed in [21], which can be subtracted from the FL-GPR measurements to reduce the clutter. A multi-view approach based on the likelihood ratio tests (LRT) detector was proposed in [14] and its adaptive counterpart was presented in [22] for effective detection of low-signature targets in the presence of rough surface clutter. A robust LRT was designed in [23] based on the least favorable target and clutter densities to maximize the worst-case detection performance over all feasible target and clutter models in FL-GPR images. In [24], infrared imagery was used to eliminate false alarms in FL-GPR.

On the other hand, a variety of image formation approaches have been considered in the context of array-based FL-GPR imaging. The most commonly used algorithms are back-projection, frequency-wavenumber migration, and scalar inverse scattering [12,25,26]. These algorithms rely on a scalar representation of the electromagnetic field, for which the relevant Green's function is simplified by assuming a free-space propagation model. Within the framework of linear inverse scattering, a two-dimensional imaging algorithm for bistatic FL-GPR systems was recently proposed in [15], whereas an inverse processing scheme that exploits the intrinsic multi-aperture nature of the FL-GPR geometry was designed in [14]. Data-adaptive approaches for FL-GPR have also been proposed in the literature [10,13]. Amplitude and phase estimation and rank-deficient robust Capon beamforming were presented in [10], while an iterative hyperparameter-free maximum a posteriori probability algorithm was proposed in [13].

In this paper, we present an FL-GPR image enhancement procedure that employs tomographic imaging and coherence-factor (CF) based masking operation for rough surface clutter suppression. More specifically, we first employ a matched-filtering (MF) based tomographic imaging approach for image formation, which exploits the vectorial nature of the incident and scattered electric fields, in conjunction with coherent combining of multiple measurements from different aperture positions. This approach builds on the MF formulation of [27] for DL-GPR imaging. Following image reconstruction, we perform a masking operation with a coherence map of the scene for clutter suppression. The coherence map is generated using the CF, which represents a measure of the relative coherence of the received signals across all antennas. We consider three variants of the CF, namely, the amplitude CF (ACF), the phase CF (PCF), and the sign CF (SCF) [28–30]. The capability of the proposed procedure to significantly suppress the clutter generated by the backscattering from a rough surface is demonstrated using near-field electromagnetic modeled numerical data corresponding to a scene with both plastic and metallic targets buried at shallow depths below a rough interface [31]. The improvements achievable are quantified in terms of the image-domain signal-to-clutter ratio (SCR), starting with the preliminary investigation reported in [32]. We show that all variants of the CF successfully suppress the rough surface clutter with comparable SCR values, and the hybrid MF-based imaging and CF-based masking procedure outperforms the case when CF masking is used in conjunction with standard back-projection (BP). It is noted that two-dimensional (2D) versions of the CF, recently proposed in [33] for sidelobe suppression in radar imaging, can also be employed in the proposed scheme. However, as our objective is to demonstrate the offerings of the hybrid procedure and not specifically identify an optimal method for coherence map generation, we will not consider the 2D versions in this paper.

The remainder of this paper is organized as follows. The various methods considered in this paper are presented in Section 2. More specifically, the MF-based tomographic imaging algorithm is described in Section 2.1, while the BP algorithm is briefly discussed in Section 2.2. The CF-based image enhancement method is presented in Section 2.3, wherein the three variants of CF and the SCR in the image domain are defined. In Section 3, we describe the considered FL-GPR configuration and simulation set up, and provide both the imaging and CF-based enhanced results. BP-based results are also presented for comparison therein. Insights into the performance of the proposed scheme are provided in Section 4. Conclusion follows in Section 5.

2. Methods

2.1. Matched-Filtering-Based Near-Field Tomographic Imaging

We consider an FL-GPR system consisting of an N_T-element linear transmit array and an N_R-element linear receive array. The transmit and receive antennas are oriented parallel to the y-axis in the yz-plane and mounted on top of a vehicle at different heights (z-coordinates). The investigation domain is located on the ground in front of the vehicle along the x-axis (see Figure 1). The transmitters are assumed to be activated sequentially, with simultaneous reception at all receivers, as the vehicle moves forward. For convenience, we assume that a single transmitter is active for each platform position. Thus, a full-aperture measurement set comprises $N_T N_R$ observations from N_T consecutive platform positions. The frequency band of operation extends from ω_l to ω_H.

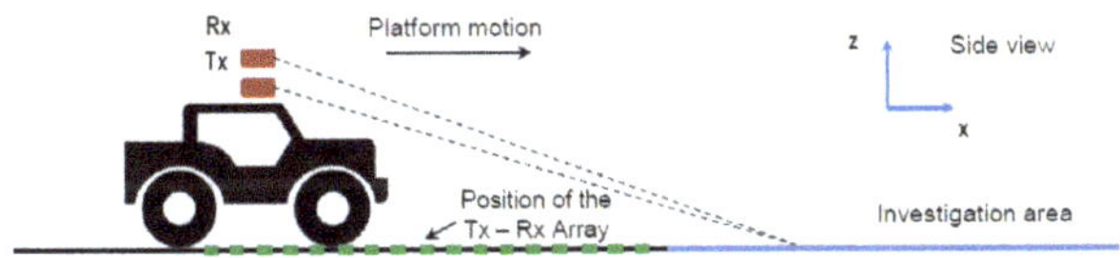

Figure 1. Side-view of the forward-looking ground-penetrating radar (FL-GPR) configuration and data collection geometry.

Considering a 3-D version of the well-known scattering equation, a linear scattering model can be established under the Born approximation for the near-field imaging conditions of the considered scenario as [34].

$$\mathbf{E}_s(\mathbf{r}_{rn}, \mathbf{r}_{tm}, \omega) = k_b^2 \iint_D \mathbf{G}(\mathbf{r}, \mathbf{r}_{rn}, \omega) \cdot \mathbf{E}_{inc}(\mathbf{r}, \mathbf{r}_{tm}, \omega) O(\mathbf{r}) d\mathbf{r}. \tag{1}$$

This model represents the relationship between the scattered field $\mathbf{E}_s$ from the investigation domain D, recorded at the n-th receive location $\mathbf{r}_{rn}$ with the m-th active transmitter at $\mathbf{r}_{tm}$, and the unknown scene reflectivity $O(\mathbf{r})$ for angular frequency ω. In (1), $\mathbf{G}$ is the dyadic Green's function of the problem, $\mathbf{E}_{inc}$ is the incident field, which under the Born approximation represents the total field inside the domain D, $k_0 = \omega\sqrt{(\varepsilon_0\mu_0)}$ is the free-space wavenumber, $k_b = \sqrt{\varepsilon_r}k_0$ is the wavenumber of the subsurface medium, and $\mathbf{r}$ represents a generic point in the domain D. The vectors $\mathbf{r}, \mathbf{r}_{tm}$, and $\mathbf{r}_{rn}$ are defined as

$$\mathbf{r}_{tm} = x_{tm}\mathbf{x}_0 + y_{tm}\mathbf{y}_0 + z_{tm}\mathbf{z}_0$$
$$\mathbf{r}_{rn} = x_{rn}\mathbf{x}_0 + y_{rn}\mathbf{y}_0 + z_{rn}\mathbf{z}_0 \tag{2}$$
$$\mathbf{r} = x\mathbf{x}_0 + y\mathbf{y}_0 + z\mathbf{z}_0,$$

with $\mathbf{x}_0$, $\mathbf{y}_0$, and $\mathbf{z}_0$ denoting the unit vectors along the x, y, and z directions, respectively. The operator $(\cdot)$ in (1) represents the dyadic product and is implemented as the typical matrix-vector product between the 3×3 matrix Green's function $\underline{\mathbf{G}}$ and the 3×1 vector incident field $\mathbf{E}_{\text{inc}}$. Equation (1) accounts for the dyadic nature of the interaction between the electric field and the probed scene.

Modeling the transmitting elements as Hertzian electric dipoles oriented along $\mathbf{z}_0$, the incident electric field can be expressed as

$$\mathbf{E}_{\text{inc}}(\mathbf{r}, \mathbf{r}_{tm}, \omega) = -j\omega\mu_0 I_0 l \underline{\mathbf{G}}(\mathbf{r}, \mathbf{r}_{tm}, \omega) \cdot \mathbf{z}_0, \tag{3}$$

where $I_0 l$ is the current moment associated with the short dipole directed along $\mathbf{z}_0$ and is assumed to be equal to 1 A·m. Therefore, (1) can be rewritten as

$$\mathbf{E}_s(\mathbf{r}_{rn}, \mathbf{r}_{tm}, \omega) = -j\omega\mu_0 k_b^2 \iint_D \underline{\mathbf{G}}(\mathbf{r}; \mathbf{r}_{rn}, \omega) \cdot [\underline{\mathbf{G}}(\mathbf{r}; \mathbf{r}_{tm}, \omega) \cdot \mathbf{z}_0] O(\mathbf{r}) d\mathbf{r}. \tag{4}$$

Under the assumptions that (i) the separation in height between the transmit and receive elements is negligible, and (ii) the targets are either on the ground surface or buried at shallow depths, the dyadic Green's function and, subsequently, the incident field can be approximated as those modeling propagation in a homogeneous medium having the electromagnetic properties of free-space [34]. That is,

$$\underline{\mathbf{G}}(\mathbf{r}, \mathbf{r}_s, \omega) = \left[\underline{\mathbf{I}} + \frac{\nabla\nabla}{k_0^2}\right] \frac{e^{-jk_0|\mathbf{r} - \mathbf{r}_s|}}{4\pi|\mathbf{r} - \mathbf{r}_s|}, \tag{5}$$

where $\underline{\mathbf{I}}$ is the unit dyad and $s = rn$ or tm.

Dividing the domain D into a finite number of pixels, say Q, we assume only one point scatterer exists per pixel. Ignoring the mutual interactions between scatterers, the point target at the q-th pixel can be modeled as an impulse located at the considered pixel, whose position vector is denoted by $\mathbf{r}_q$. As a result, the scattered field from the q-th image pixel recorded by the n-th receiver with the m-th transmitter active and directed along $\mathbf{z}_0$ is given by

$$\mathbf{E}_s(\mathbf{r}_{rn}, \mathbf{r}_{tm}, \omega) = -j\omega\mu_0 k_0^2 [\underline{\mathbf{G}}(\mathbf{r}_q, \mathbf{r}_{rn}, \omega)] \cdot [\underline{\mathbf{G}}(\mathbf{r}_q; \mathbf{r}_{tm}, \omega) \cdot \mathbf{z}_0] O(\mathbf{r}_q). \tag{6}$$

If only the z-component of the electric field is measured by the receiving antenna (i.e., through a linear polarized receiving antenna modeled as a short dipole oriented along $\mathbf{z}_0$), we can express the recorded electric field E_{s_z} as

$$E_{s_z}(\mathbf{r}_{rn}, \mathbf{r}_{tm}, \omega) = \mathbf{z}_0 \cdot \mathbf{E}_s(\mathbf{r}_{rn}, \mathbf{r}_{tm}, \omega) = -j\omega\mu_0 k_0^2 \left(G_{zx}G_{xz} + G_{zy}G_{yz} + G_{zz}G_{zz}\right) O(\mathbf{r}_q), \tag{7}$$

where the Green's functions components, G_{ij}, with i and j representing the Cartesian coordinates x, y, z, can be derived from (5) [5].

With both transmitting and receiving antennas linearly polarized along the z-axis, we define the frequency response $H_{zz}(\mathbf{r}_{rn}, \mathbf{r}_{tm}, \mathbf{r}_q, \omega)$ of a filter matched to a point scatterer with unit reflectivity at pixel $\mathbf{r}_q$, when the m-th antenna is transmitting and the n-th antenna is receiving, using (7) as

$$H_{zz}(\mathbf{r}_{rn}, \mathbf{r}_{tm}, \mathbf{r}_q, \omega) = \left\{-j\omega\mu_0 k_0^2 \left(G_{zx}G_{xz} + G_{zy}G_{yz} + G_{zz}G_{zz}\right)\right\}^*, \tag{8}$$

with '$*$' denoting complex conjugation. The reflectivity estimate $\hat{O}_{nm}(\mathbf{r}_q)$ of the q-th pixel is obtained by applying the matched filter to the recorded measurements by the n-th receiver when the m-th antenna is transmitting over the bandwidth of interest as

$$\hat{O}_{nm}(\mathbf{r}_q) = \int_{\omega_L}^{\omega_H} H_{zz}(\mathbf{r}_q, \mathbf{r}_{tm}, \mathbf{r}_{rn}, \omega) E_{sz}(\mathbf{r}_{rn}, \mathbf{r}_{tm}, \omega) d\omega. \tag{9}$$

Note that (9) provides the reflectivity estimate for each pixel as a function of the transmitter and receiver locations. The reflectivity estimate for the pixel at $\mathbf{r}_q$, corresponding to all N_T transmitting and N_R receiving z-polarized antennas, can be obtained by exploiting (8) and (9) as,

$$\begin{aligned} \hat{O}(\mathbf{r}_q) &= \sum_{n=1}^{N_R} \sum_{m=1}^{N_T} \hat{O}_{nm}(\mathbf{r}_q) \\ &= j\omega\mu_0 k_0^2 \sum_{n=1}^{N_R} \sum_{m=1}^{N_T} \int_{\omega_L}^{\omega_H} \left(G_{zx}G_{xz} + G_{zy}G_{yz} + G_{zz}G_{zz} \right)^* E_{s_z}(\mathbf{r}_{rn}, \mathbf{r}_{tm}, \omega) d\omega. \end{aligned} \tag{10}$$

The spatial map, $\hat{O}(\mathbf{r}_q)$ of the scene reflectivity is the desired image of the investigated domain D and is the final outcome of the MF-based imaging algorithm. It is noted that coherent integration of measurements corresponding to multiple full apertures, resulting from radar platform motion, can also be employed within the MF imaging framework to reduce artifacts and rough surface clutter prior to the CF-based masking operation [14,18].

2.2. Back-Projection Algorithm

An alternative approach to generating FL-GPR images is the BP algorithm, which is based on scalar wave theory [1]. The mathematical formulation for the BP-based image formation method can be essentially derived by simplifying the dyadic Green's function in (5) to a scalar model. Thus, the reflectivity estimate at pixel location $\mathbf{r}_q$, assuming a free-space propagation model, is achieved as

$$\hat{O}^{BP}(\mathbf{r}_q) = \sum_{n=1}^{N_R} \sum_{m=1}^{N_T} \hat{O}_{nm}^{BP}(\mathbf{r}_q) = \sum_{n=1}^{N_R} \sum_{m=1}^{N_T} \int_{\omega_L}^{\omega_H} e^{jk_0|\mathbf{r}_q - \mathbf{r}_{tm}|} e^{jk_0|\mathbf{r}_q - \mathbf{r}_{rn}|} E_{s_z}(\mathbf{r}_{rn}, \mathbf{r}_{tm}, \omega) d\omega. \tag{11}$$

The spatial map, $\hat{O}^{BP}(\mathbf{r}_q)$, of the scene reflectivity represents the BP-based image of the investigated domain D.

2.3. Coherence-Factor-Based Image Enhancement

In this section, we present the CF-based processing for the enhancement of cluttered FL-GPR images. We consider the following three variants of the CF: the amplitude CF (ACF), the phase CF (PCF), and the sign CF (SCF).

The ACF is defined as the ratio of the total coherent power received by the antenna array (generated by the presence of targets in the domain under investigation) to the total incoherent power (produced by the rough surface clutter for the case under consideration). Mathematically, it can be expressed as [32]

$$\text{ACF}(\mathbf{r}_q) = \frac{\left| \sum_{n=1}^{N_R} \sum_{m=1}^{N_T} \hat{O}_{nm}(\mathbf{r}_q) \right|^2}{N_R N_T \sum_{n=1}^{N_R} \sum_{m=1}^{N_T} \left| \hat{O}_{nm}(\mathbf{r}_q) \right|^2}, \tag{12}$$

with $\hat{O}_{nm}$ given by (9) and $N_R N_T$ representing the total number of receive channels in the full aperture. From (12), it follows that the ACF varies from zero to unity. It assumes small values for low-coherence image regions corresponding to rough surface clutter and high values for target regions. As such, the coherence map of the scene, generated by computing (12) for all Q pixels, can be used to perform a corrective action on the MF-based image, $\hat{O}(\cdot)$, as

$$\hat{O}_{CF}(\mathbf{r}_q) = ACF(\mathbf{r}_q)\hat{O}(\mathbf{r}_q). \tag{13}$$

That is, the enhanced image is the pixel-by-pixel multiplication of the coherence map, defined by (12), times the output of the MF-based tomographic algorithm in (10). Clearly, low-coherence rough surface clutter will be suppressed or significantly attenuated.

Unlike the ACF, the PCF exploits the phase disparity across the antenna array [35,36]. It is defined as

$$PCF(\mathbf{r}_q) = 1 - std(e^{j\angle\hat{O}(\mathbf{r}_q)}), \tag{14}$$

where $\angle\hat{\mathbf{O}}(\mathbf{r}_q) = \{\angle\hat{O}_{nm}(\mathbf{r}_q), n = 1,\ldots,N_R, m = 1,\ldots,N_T\}$ and "std" denotes the standard deviation of the complex exponential term. The PCF corrected image is obtained through (13) by simply replacing $ACF(\mathbf{r}_q)$ with $PCF(\mathbf{r}_q)$ defined by (14).

The SCF can be derived from the PCF by introducing a sign bit as follows [37]. The pixel phase $\angle\hat{O}_{nm}$ is quantized with a single bit, thereby splitting the interval $[-\pi, \pi]$ in two sub-intervals, namely, $(-\pi/2, \pi/2]$ and $[-\pi, -\pi/2] \cup (\pi/2, \pi]$, and the sign bit b_{nm} is obtained as,

$$b_{nm}(\mathbf{r}_q) = \begin{cases} -1, & real(\hat{O}_{nm}(\mathbf{r}_q)) < 0 \\ +1 & real(\hat{O}_{nm}(\mathbf{r}_q)) \geq 0 \end{cases}. \tag{15}$$

The SCF can then be defined as,

$$SCF(\mathbf{r}_q) = 1 - std(\mathbf{b}_q), \tag{16}$$

where $\mathbf{b}_q = \{b_{nm}(\mathbf{r}_q), n = 1,\ldots,N_R, m = 1,\ldots,N_T\}$. Again, the SCF corrected image is obtained using (13) by substituting $CF(\mathbf{r}_q)$ with $SCF(\mathbf{r}_q)$.

We note that the CF-based correction, proposed for enhancing images obtained with the MF-based tomographic algorithm, can also be applied to images generated using the BP algorithm in (11); the coherence map, generated using any variant of the CF, will also then be based on the BP approach. That is, the pixel values $\hat{O}_{nm}(\mathbf{r}_q)$ in (12)–(16) will be replaced with the corresponding values of the back-projected image. It is important to note that applying the definitions of the ACF, PCF, and SCF, as presented in (12)–(16), to MF-based imaging provides enhanced imaging compared to the case when these coherence factors are applied to the BP-based imaging. The former exploits the vector nature of the scattering mechanism unlike the latter. Thus, the proposed hybrid MF-based imaging and CF-based masking provides a two-fold advantage in terms of modeling accuracy over its BP-based counterpart.

In order to obtain a quantitative assessment of the image enhancements offered by the CF-based procedure, we employ the image-domain SCR as a metric [29,38]. The SCR is defined as the ratio of the average amplitude of the pixels associated with the targets in the enhanced image to the average of those related to clutter. That is,

$$SCR = 10\log_{10}\left[\frac{\frac{1}{N}\sum_{\mathbf{r}_q\in\mathcal{R}_t}|\hat{O}_{CF}(\mathbf{r}_q)|^2}{\frac{1}{M}\sum_{\mathbf{r}_q\in\mathcal{R}_c}|\hat{O}_{CF}(\mathbf{r}_q)|^2}\right], \tag{17}$$

where N and M denote the respective number of pixels in the target region $\mathscr{R}_t$ and the clutter region $\mathscr{R}_c$. A region growing algorithm can be used to isolate the targets comprising $\mathscr{R}_t$ [39], and the remainder of the image constitutes $\mathscr{R}_c$.

3. Results

In this section, we describe the electromagnetic simulation set up and then present CF-based image enhancement results which demonstrate the capability of the proposed procedure to suppress rough surface clutter in FL-GPR images.

3.1. Radar Configuration

A stepped-frequency multi-antenna FL-GPR, mounted on top of a vehicle, is modeled in AFDTD, which is a full-wave near-field electromagnetic software based on a finite-difference time-domain (FDTD) algorithm [18,31]. The radar system operates over the 0.3–1.5 GHz frequency band, with a forward-looking coverage angle spanning approximately 5°–20° with respect to the horizon. Two transverse electromagnetic horn antennas are used as transmitters, whose near-field configuration is represented by an equivalent current distribution. In between the two transmitters are the 16 uniformly-spaced receiving short-dipole antennas. Both the transmit and receive antennas are distributed over a 2-m wide aperture and are placed 2 m and 1.9 m above the rough ground surface, respectively. The radar system parameters are summarized in Table 1.

Table 1. Main characteristics of the radar system and the investigation area.

Investigation area	Size: 10×16 m	$\varepsilon_r = 6; \sigma_d = 10$	h_{rms} 1.8 cm , $l = 14.26$ cm
Antenna height	Tx antennas: 1.9 m	Rx antennas: 2 m	
Linear antenna array	Aperture extent: 2 m	Rx antennas: 16	Tx antennas: 2
System parameters	Frequency: 0.3–1.5 GHz	Coverage angle: 5°–20°	

For each position of the moving platform, only one of the two transmitters is activated. By alternating between the left and right antennas from one platform position to the next, a full aperture comprising 32 receive channels is obtained from two consecutive platform positions or scans. The first position of the system on the surface is at $x = -12$ m and the last one at $x = 11$ m, as shown in Figure 2. Since the system radiates and collects data along the x-direction with a discretized step of $\delta_x = 0.33$ m, we have a total of 70 scans (represented in Figure 2 with black vertical lines). The ground is modeled as a non-dispersive and non-magnetic homogeneous medium with effective relative dielectric constant $\varepsilon_r = 6$ and conductivity $\sigma_d = 10$ mS/m. A rough profile for the interface separating the upper and lower dielectric half-spaces is introduced and a statistical model is exploited to provide a realistic representation in the numerical code. The model is described by two functions [18]: the probability density function of the height variations and the surface autocorrelation function. For the numerical data considered in this paper, a 2-D zero-mean surface profile represented by Gaussian statistics (described by two parameters: the *rms* height h_{rms} and the correlation length l_c) is assumed. Thus, the scattered electric field is considered to be a random process and evaluated by means of a Monte Carlo simulation [18,40].

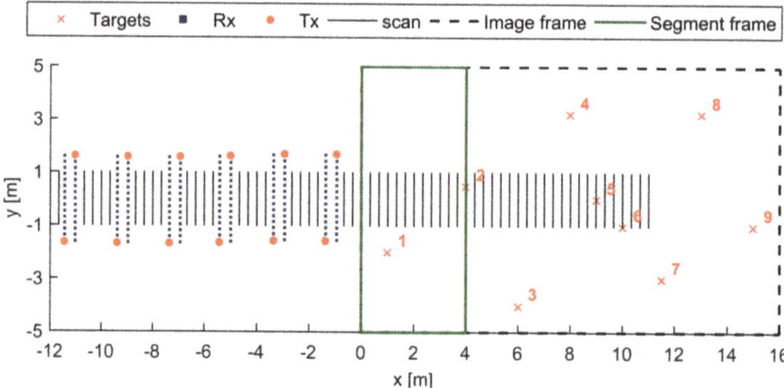

Figure 2. Top view of the numerical simulation geometry. Six different full-aperture measurements have been highlighted in blue, which are used in coherent combining for the first image segment indicated in green (the positions of transmitters and receivers are not drawn to scale within the image frame).

The investigation area (indicated by a black dashed rectangle in Figure 2 has dimensions of 10 m × 16 m along y and x directions, respectively, and is populated by a total of nine targets at distinct locations. The target characteristics are summarized in Table 2.

Table 2. Target characteristics.

Target No.	Type	State	Size
1	Metallic anti-personnel landmine	Buried	Diameter: 100 mm; Height: 55 mm
2	Plastic anti-personnel landmine	On surface	Diameter: 100 mm; Height: 55 mm
3	Metallic artillery shell	Buried	Diameter: 155 mm; Length: 585 mm
4	Metallic anti-tank landmine	Buried	Diameter: 300 mm; Height: 125 mm
5	Metallic anti-tank landmine	On surface	Diameter: 300 mm; Height: 125 mm
6	Metallic artillery shell	Buried	Diameter: 155 mm; Length: 585 mm
7	Metallic artillery shell	Buried	Diameter: 155 mm; Length: 585 mm
8	Plastic anti-personnel landmine	On surface	Diameter: 100 mm; Height: 55 mm
9	Plastic anti-tank landmine	Buried	Diameter: 300 mm; Height: 125 mm

Buried targets are positioned 3 cm below the surface. The plastic targets have a relative dielectric constant $\varepsilon_r = 3.1$ and conductivity $\sigma = 2$ mS/m. For the rough ground surface, $h_{\mathrm{rms}} = 1.6$ cm and $l_c = 14.26$ cm. The characteristics of the investigation area are summarized in Table 1.

3.2. Image Formation Results

In order to maintain a similar cross-range resolution over the entire image, the investigation area is divided into four segments, each of dimension 10 m × 4 m. The first segment is highlighted in Figure 2 with a green rectangle. Since coherent integration has been shown to reduce clutter [31], we coherently add multiple images of each segment generated with measurements from a set of full apertures using the MF-based tomographic algorithm detailed in Section 2.1. The CF-based processing of Section 2.3 is then applied to the resulting composite image. The set of apertures for each segment are selected so that the standoff distances are the same across all the image segments. Instead of choosing consecutive apertures for each segment, we opt for a set of apertures wherein any two neighboring apertures are separated by $4\delta_x = 4(0.33) = 1.32$ m, with the aperture closest to the segment at a standoff distance of 1 m. Figures 2 and 3 depict the respective sets of full apertures used for the first and the last segments (indicated as blue dashed vertical lines). Such a choice provides a larger variation of the clutter across the various images being combined. For more details on the coherent combining procedure, see [14].

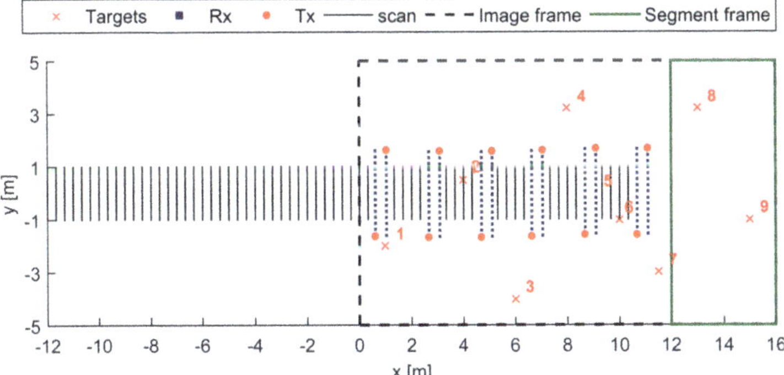

Figure 3. The last image segment is highlighted in green and the relevant full aperture measurements to be exploited in coherent combining are indicated in blue (the positions of transmitters and receivers are not drawn to scale within the image frame).

In Figure 4, we present the MF-based composite image corresponding to two full apertures (ones closest to each segment), whereas that corresponding to six full apertures is depicted in Figure 5. These results and all subsequent images in this paper are plotted on a 40 dB dynamic range, unless otherwise stated, with the maximum intensity value in each image normalized to 0 dB. The target positions are indicated with white crosses in both Figures 4 and 5. The clutter generated by the rough surface dominates the image in Figure 4 and obscures the low-signature targets. Owing to the integration of a larger number of apertures permitted by the considered FL-GPR configuration, the clutter in Figure 5 is reduced as compared to Figure 4. Nonetheless, there is still substantial residual clutter in Figure 5, which would render target detection challenging. This demonstrates the need for further enhancements via the proposed CF procedure.

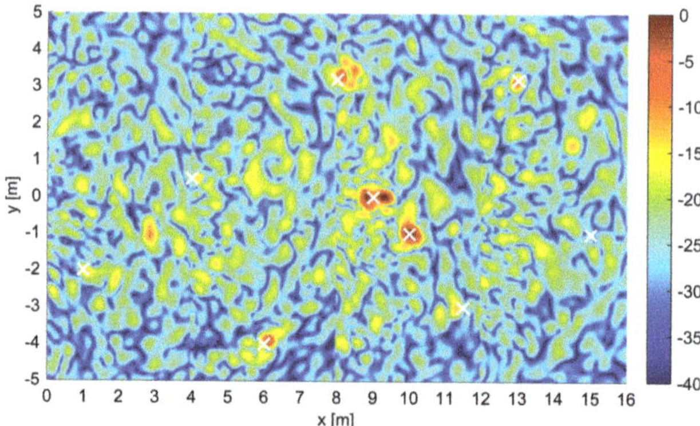

Figure 4. Image obtained by integrating two apertures for the scene containing nine targets and a rough surface ($h_{rms} = 1.6$ cm). The true position of targets is indicated with a white cross.

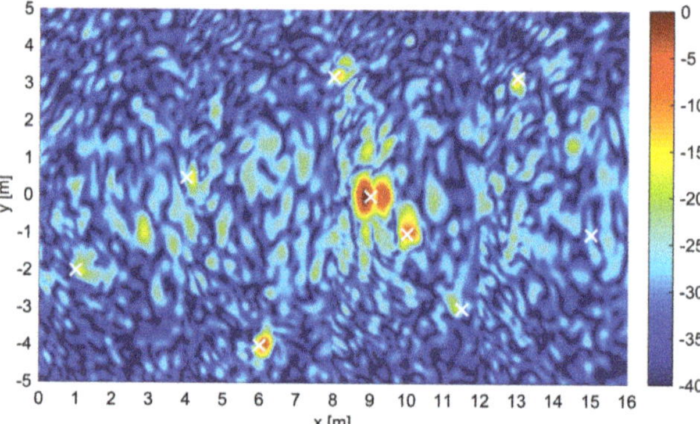

Figure 5. Image obtained by integrating six apertures for the same scene as in Figure 4.

For comparison, we provide in Figures 6 and 7 the images obtained by exploiting the same apertures as in Figures 4 and 5, but with a BP algorithm [1]. As expected, the coherent combining of six full apertures allows for higher clutter suppression. Comparing the MF-based images with their respective BP counterparts, the improvements offered by the more accurate vector model adopted by the MF tomographic algorithm over the scalar-model-based BP algorithm are clearly visible in the central part of the images, where the clutter manifests itself as relatively weaker in strength. These qualitative observations are also validated by the corresponding SCR values, listed in Table 3. More specifically, the MF algorithm provides an improvement of 1.3 dB and 2.9 dB over the BP algorithm for the 2- and 6-apertures cases, respectively.

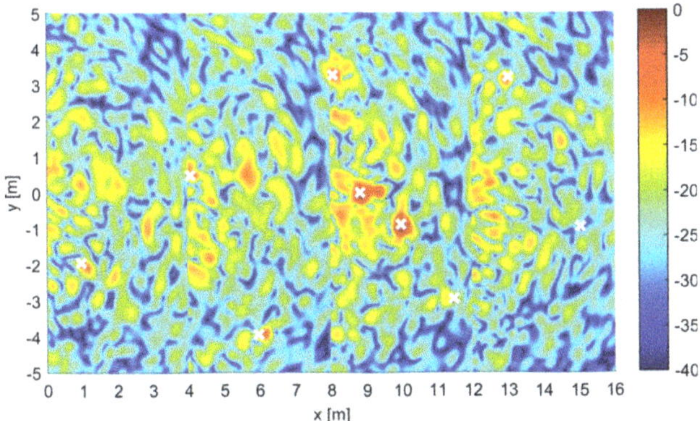

Figure 6. Image as in Figure 4, but generated through the back-projection (BP) algorithm.

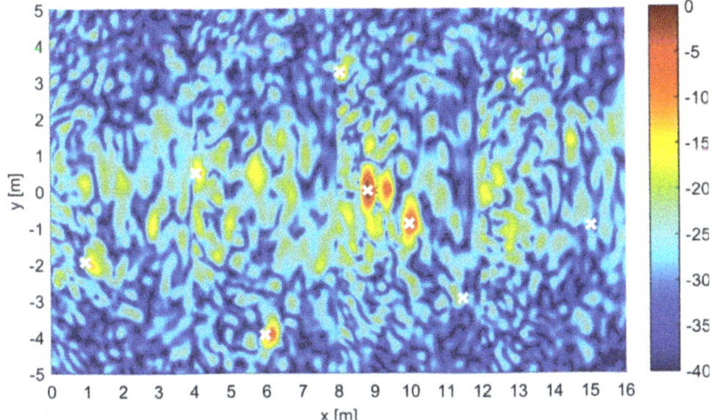

Figure 7. Image as in Figure 5, but obtained by means of the BP algorithm.

Table 3. Signal-to-clutter ratio (SCR) for back-projected and matched-filtering (MF)-based tomographic images.

	Figure Number	SCR (dB)
MF 2 apertures	Figure 4	−7.4
MF 6 apertures	Figure 5	−0.4
BP 2 apertures	Figure 6	−8.7
BP 6 apertures	Figure 7	−3.3

3.3. CF Enhanced Results

We first apply the enhancement procedure based on ACF to both MF and BP images, and demonstrate the superior clutter suppression capability yielded by the MF-based ACF over that defined using the scalar-model-based BP algorithm.

Figure 8 depicts the image obtained by means of the ACF based masking operation applied to the two-aperture MF image of Figure 4. The image enhancements in terms of clutter mitigation are clearly visible with respect to the original. Figure 9 shows the two-aperture BP image of Figure 6 after the BP-based ACF masking operation was applied. Comparing Figures 8 and 9, we observe that the

MF-based enhancement procedure provides a higher degree of clutter suppression. The six-aperture MF and BP images after application of the ACF-based correction are shown in Figures 10 and 11, respectively. As expected, more clutter has been suppressed with respect to the two-aperture configuration for both cases. Similar to the two-apertures case, the MF-based definition of the ACF provides a cleaner image, which would lead to an improved detection performance.

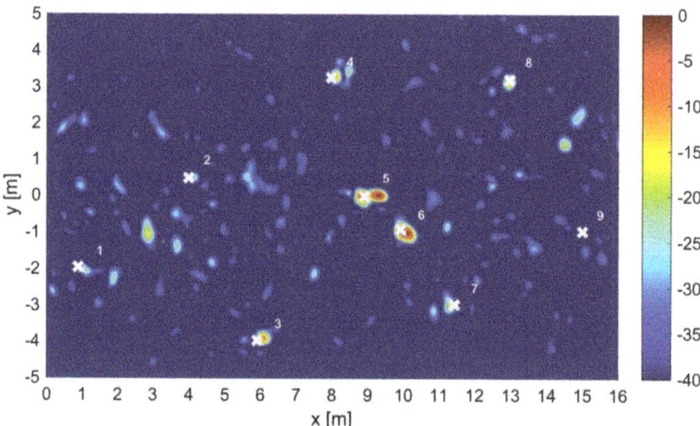

Figure 8. MF based image of Figure 4 after application of the ACF-based enhancement procedure. The target position and type are indicated with a white cross and number.

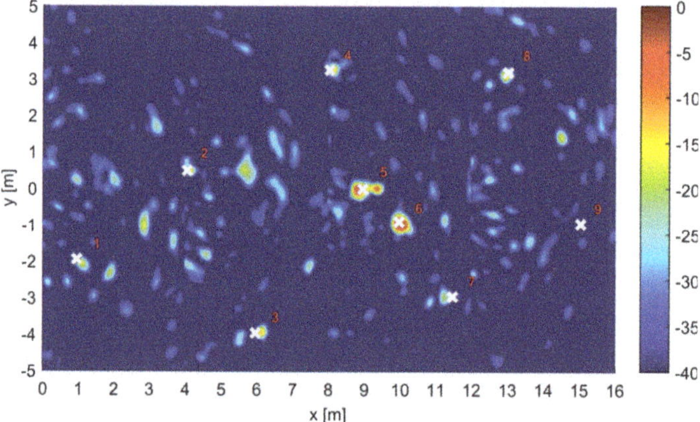

Figure 9. BP image of Figure 6 after application of the amplitude coherence factor (ACF)-based enhancement procedure. The target position and type are indicated with a white cross and number.

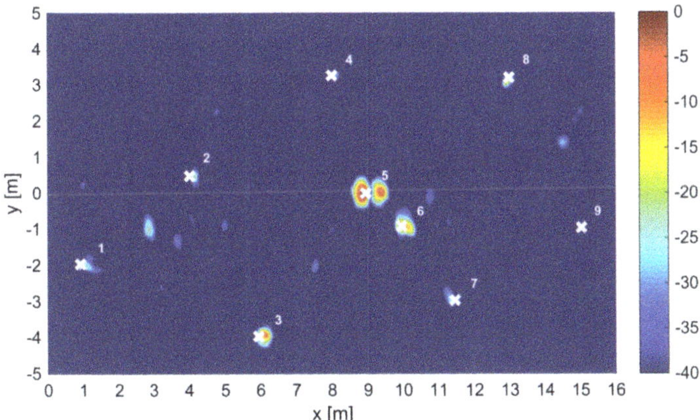

Figure 10. MF image of Figure 5 after ACF based enhancement.

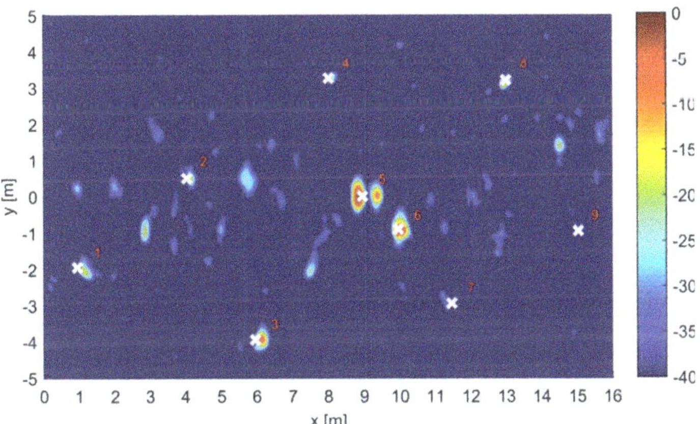

Figure 11. BP image of Figure 7 after ACF based enhancement.

Having demonstrated the superiority of the MF-based proposed procedure over the BP-based enhancement, we next compare and contrast the performance of the CF-based scheme when ACF, PCF, and SCF are individually used to generate the coherence maps for clutter suppression in MF images. We consider the MF image of Figure 5 (six-aperture case) for this purpose. Figures 12 and 13 present the resulting images after application of the clutter suppression procedure via PCF and SCF, respectively. Comparing Figure 12 and Figure 13 with the ACF corrected image of Figure 10, we observe that the different coherence map definitions provide comparable degree of clutter suppression. This is also demonstrated by the corresponding SCR values, provided in Table 4. More specifically, all three variants of CF provide SCR improvements of 7 to 8 dB over the original image of Figure 5. Similar results were obtained when the three variants of the CF were applied to the MF image obtained through the coherent combining of two apertures.

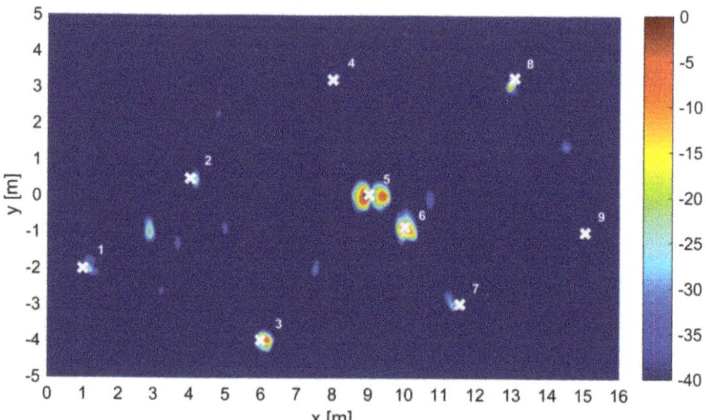

Figure 12. MF Image of Figure 5 corrected via phase coherence factor (PCF).

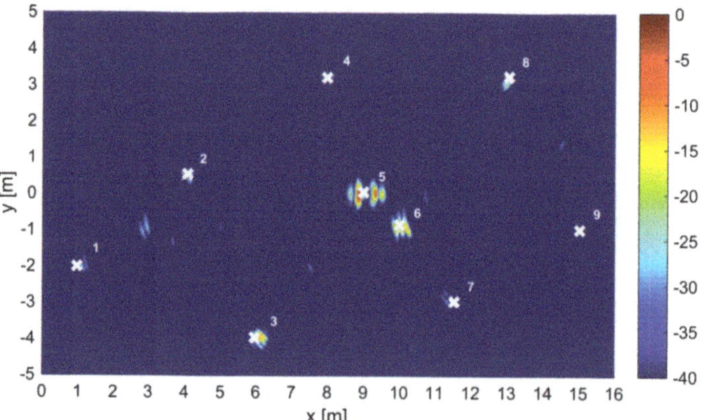

Figure 13. MF image of Figure 5 corrected via sign coherence factor (SCF).

Table 4. SCR for six-aperture images after CF-based enhancement.

	Figure Number	SCR (dB)
ACF MF	Figure 10	6.7
ACF BP	Figure 11	5.6
PCF MF	Figure 12	7.0
SCF MF	Figure 13	7.6

4. Discussion

The qualitative and quantitative results of Section 3 clearly demonstrated the superior performance of the CF clutter suppression approach based on MF image formation over its BP-based counterpart. This superiority is attributed to the high-accuracy vector model employed by the MF algorithm over the scalar-model-based BP algorithm. Further, coherent integration of measurements from multiple full apertures should be employed, whenever possible, in conjunction with the CF-based approach for a higher degree of clutter suppression. Furthermore, performance evaluation of different coherence map definitions, namely, ACF, PCF, and SCF, showed that all three variants of CF provide comparable levels of clutter suppression. In terms of the impact of the CF-based processing on the target regions, we observed that both ACF and PCF had a minimal effect, as evident from Figures 10 and 12. However, each target region in the SCF-corrected image split up into multiple lobes, as evident in Figure 13, which may be problematic for subsequent target detection schemes. Finally, we note that target 9 did not survive the clutter suppression process and was missing from all CF-corrected results reported in Section 3. This is because the target in question is the only plastic target buried in the ground. Buried plastic targets are especially hard to detect due to (i) the limited dielectric contrast between the target and the soil background, and (ii) interference from rough surface scattering. This observation is consistent with what has been previously reported in the literature [31].

5. Conclusions

In this paper, we proposed a matched filtering formulation of tomographic near-field imaging and presented a coherence-factor-based rough surface clutter mitigation technique for FL-GPR imaging. The CF was used to generate a coherence map, which was then applied as a correction mask to the microwave image. Improvements achievable, in terms of reduction of the incoherent component produced by the rough surface with respect to coherent scattering from targets, were assessed using numerical data of metallic and plastic targets both on-surface and buried at shallow depths. The performance of the proposed scheme was also quantified by evaluating the improvements in image-domain SCR and contrasted with that obtained using a standard back-projection imaging algorithm. The proposed approach was shown to outperform the back-projection-based scheme. Different definitions of the CF were considered and compared. It was shown that the three variants of the CF all yielded comparable but excellent SCR enhancements. While the SCF generated some artifacts by splitting each target into multiple lobes, both the ACF and PCF exhibited minimal impact on the weak target signatures.

Author Contributions: Conceptualization, F.A. and M.A.; methodology and formal analysis, D.C. and F.A.; data generation: T.D.; validation, D.C.; writing—original draft preparation, D.C.; writing—review and editing, F.A., M.A., and T.D. All authors have read and agreed to the published version of the manuscript.

Funding: This work was supported by the US Army Research Office and US Army Research Lab under contract W911NF-11-1-0536.

Conflicts of Interest: The authors declare no conflict of interest.

References

1. Pastorino, M. *Microwave Imaging*; John Wiley & Sons Inc.: Hoboken, NJ, USA, 2010.
2. Jin, S.; Haghighipour, N.; Ip, W.-H. (Eds.) *Planetary Exploration and Science: Recent Results and Advances*; Springer: Berlin/Heidelberg, Germany, 2015.
3. Amin, M.G. *Through the-Wall Radar Imaging*; CRC Press: Boca Raton, FL, USA, 2011.
4. Solimene, R.; Catapano, I.; Gennarelli, G.; Cuccaro, A.; Dell Aversano, A.; Soldovieri, F. SAR imaging algorithms and some unconventional applications. *IEEE Signal Process. Mag.* **2014**, *31*, 90–98. [CrossRef]
5. Persico, R. *Introduction to Ground Penetrating Radar: Inverse Scattering and Data Processing*; John Wiley & Sons Inc.: Hoboken, NJ, USA, 2014.
6. Lo Monte, L.; Erricolo, D.; Soldovieri, F.; Wicks, M.C. Radio frequency tomography for tunnel detection. *IEEE Trans. Geosci. Remote Sens.* **2010**, *48*, 1128–1137. [CrossRef]
7. Garcia-Fernandez, M.; Morgenthaler, A.; Alvarez-Lopez, Y.; Las Heras, F.; Rappaport, C. Bistatic landmine and IED detection combining vehicle and drone mounted GPR sensors. *Remote Sens.* **2019**, *11*, 2299. [CrossRef]

8.   Kositsky, J.; Cosgrove, R.; Amazeen, C.; Milanfar, P. Results from a forward-looking GPR mine detection system. *Proc. SPIE* **2002**, *4742*, 206–217.

9.   Kositsky, E.M.; Rotondo, F.S.; Elizabeth, A. Testing and evaluation of forward-looking GPR countermine systems. *Proc. SPIE* **2005**, *5794*, 901–911.

10.   Wang, Y.; Sun, Y.; Li, J.; Stoica, P. Adaptive imaging for forward-looking ground penetrating radar. *IEEE Trans. Aerosp. Electron. Syst.* **2005**, *41*, 922–936. [CrossRef]

11.   Wang, T.; Keller, J.M.; Gader, P.D.; Sjahputera, O. Frequency sub-band processing and feature analysis of forward-looking ground penetrating radar signals for land-mine detection. *IEEE Trans. Geosci. Remote Sens.* **2007**, *45*, 718–729. [CrossRef]

12.   Soldovieri, F.; Gennarelli, G.; Catapano, I.; Liao, D.; Dogaru, T. Forward-looking radar imaging: A comparison of two data processing strategies. *IEEE J. Sel. Top. Appl. Earth Obs. Remote Sens.* **2017**, *10*, 562–571. [CrossRef]

13.   Ogworonjo, H.C.; Anderson, J.M.M.; Nguyen, L.H. An iterative parameter-free MAP algorithm with an application to forward looking GPR imaging. *IEEE Trans. Geosci. Remote Sens.* **2017**, *55*, 1573–1586. [CrossRef]

14.   Comite, D.; Ahmad, F.; Liao, D.; Dogaru, T.; Amin, M.G. Multiview imaging for low-signature target detection in rough-surface clutter environment. *IEEE Trans. Geosci. Remote Sens.* **2017**, *55*, 5520–5229. [CrossRef]

15.   Catapano, I.; Affinito, A.; Del Moro, A.; Alli, G.; Soldovieri, F. Forward-looking ground-penetrating radar via a linear inverse scattering approach. *IEEE Trans. Geosci. Remote Sens.* **2015**, *53*, 5624–5633. [CrossRef]

16.   Sun, Y.; Li, J. Time-frequency analysis for plastic landmine detection via forward-looking ground penetrating radar. *Proc. Inst. Elect. Eng. Radar Sonar Navig.* **2003**, *150*, 253–261. [CrossRef]

17.   Nguyen, L. SAR imaging technique for reduction of sidelobes and noise. *Proc. SPIE* **2009**, *7308*, 73080U.

18.   Liao, D.; Dogaru, T.; Sullivan, A. Large-scale, full-wave-based emulation of step-frequency forward-looking radar imaging in rough terrain environments. *Sens. Imag.* **2014**, *15*, 88. [CrossRef]

19.   Ton, T.; Wong, D.; Soumekh, M. ALARIC forward-looking ground penetrating radar system with standoff capability. In Proceedings of the 2010 IEEE International Conference on Wireless Information Technology and Systems, Honolulu, HI, USA, 8 August–3 September 2010.

20.   Yang, J.; Jin, T.; Huang, X.; Thompson, J.; Zhou, Z. Sparse MIMO array forward-looking GPR imaging based on compressed sensing in clutter environment. *IEEE Trans. Geosci. Remote Sens.* **2014**, *52*, 4480–4494. [CrossRef]

21.   Tajdini, M.M.; Gonzalez-Valdes, B.; Martinez-Lorenzo, J.A.; Morgenthaler, A.W.; Rappaport, C.M. Real-time modeling of forward-looking synthetic aperture ground penetrating radar scattering from rough terrain. *IEEE Trans. Geosci. Remote Sens.* **2019**, *57*, 2754–2765. [CrossRef]

22.   Comite, D.; Ahmad, F.; Dogaru, T.; Amin, M.G. Adaptive detection of low-signature targets in forward-looking GPR imagery. *IEEE Geosci. Remote Sens. Lett.* **2018**, *15*, 1520–1524. [CrossRef]

23.   Pambudi, A.D.; Fauss, M.; Ahmad, F.; Zoubir, A.M. Robust detection for forward-looking GPR in rough-surface clutter environments. In Proceedings of the 52nd Asilomar Conference on Signals, Systems, and Computers, Pacific Grove, CA, USA, 28–31 October 2018; pp. 2077–2080.

24.   Keller, S.K.; Ho, K.C.; Busch, M.; Gader, P.D. On the registration of FLGPR and IR data for a forward-looking landmine detection system and its use in eliminating FLGPR false alarms. *Proc. SPIE* **2008**, *6953*, 695314.

25.   Li, L.; Zhang, W.; Li, F. Derivation and discussion of the SAR migration algorithm within inverse scattering problem: Theoretical analysis. *IEEE Trans. Geosci. Remote Sens.* **2010**, *48*, 415–422.

26.   Gilmore, C.; Jeffrey, I.; Lo Vetri, J. Derivation and comparison of SAR and frequency-wavenumber migration within a common inverse scalar wave problem formulation. *IEEE Trans. Geosci. Remote Sens.* **2006**, *44*, 1454–1461. [CrossRef]

27.   Leuschen, C.J.; Plumb, R.G. A matched-filter-based reverse-time migration algorithm for ground-penetrating radar data. *IEEE Trans. Geosci. Remote Sens.* **2001**, *39*, 929–936. [CrossRef]

28.   Hollman, K.W.; Rigby, K.W.; O Donnell, M. Coherence factor of speckle from a multi-row probe. In Proceedings of the 1999 IEEE Ultrasonics Symposium, Caesars Tahoe, NV, USA, 17–20 October 1999; pp. 1257–1260.

29.   Klemm, M.; Leendertz, J.A.; Gibbins, D.; Craddock, I.J.; Preece, A.; Benjamin, R. Microwave radar-based breast cancer detection: Imaging in inhomogeneous breast phantoms. *IEEE Antennas Wireless. Prop. Lett.* **2009**, *8*, 1949–1952. [CrossRef]

30. Burkholder, R.J.; Browne, K.E. Coherence factor enhancement of through-wall radar images. *IEEE Antennas Wirel. Prop. Lett.* **2010**, *9*, 842–845. [CrossRef]

31. Liao, D.; Dogaru, T. Full-wave characterization of rough terrain surface scattering for forward-looking radar applications. *IEEE Trans. Antennas Propag.* **2012**, *60*, 3853–3866. [CrossRef]

32. Comite, D.; Ahmad, F.; Dogaru, T.; Amin, M.G. Coherence factor for rough surface clutter mitigation in forward-looking GPR. In Proceedings of the 2017 IEEE Radar Conference (RadarConf), Seattle, WA, USA, 8–12 May 2017; pp. 1803–1806.

33. Li, S.; Amin, M.; An, Q.; Zhao, G.; Sun, H. 2-D coherence factor for sidelobe and ghost suppressions in radar imaging. *IEEE Trans. Antennas Propag.* **2019**, *68*, 1204–1209. [CrossRef]

34. Chew, W.C. *Waves and Fields in Inhomogeneous Media*; Wiley-IEEE Press: Hoboken, NJ, USA, 1995.

35. Lu, B.Y.; Sun, X.; Zhao, Y.; Zhou, Z.M. Phase coherence factor for mitigation of sidelobe artifacts in through-the-wall radar imaging. *J. Electromagn. Waves Appl.* **2013**, *27*, 716–725. [CrossRef]

36. Camacho, J.; Parrilla, M.; Fritsch, C. Phase coherence imaging. *IEEE Trans. Ultrason. Ferroelectr. Freq. Control* **2009**, *56*, 958–974. [CrossRef]

37. Liu, J.; Jia, Y.; Kong, L.; Yang, X.; Liu, Q.H. Sign-coherence-factor-based suppression for grating lobes in through-wall radar imaging. *IEEE Geosci. Remote Sens. Lett.* **2016**, *13*, 1681–1685. [CrossRef]

38. Yoon, Y.S.; Amin, M.G. Spatial filtering for wall-clutter mitigation in through-the-wall radar imaging. *IEEE Trans. Geosci. Remote Sens.* **2009**, *47*, 3192–3208. [CrossRef]

39. Adams, R.; Bischof, L. Seeded region growing. *IEEE Trans. Pattern Anal. Mach. Intell.* **1994**, *16*, 64–647. [CrossRef]

40. Chan, C.H.; Tsang, L. Monte Carlo simulations of large-scale one-dimensional random rough-surface scattering at near-grazing incidence: Penetrable case. *IEEE Trans. Antennas Propag.* **1998**, *46*, 142–149. [CrossRef]

 *remote sensing*

Article

A Compressive Sensing-Based Approach to Reconstructing Regolith Structure from Lunar Penetrating Radar Data at the Chang'E-3 Landing Site

Kun Wang [1,2]**, Zhaofa Zeng** [1,2,]*****, **Ling Zhang** [1,2]**, Shugao Xia** [3] **and Jing Li** [1,2]

[1] College of Geo-Exploration Science and Technology, Jilin University, Changchun 130026, China; wangkun0823@mails.jlu.edu.cn (K.W.); lingzhang16@mails.jlu.edu.cn (L.Z.); inter_lijing@jlu.edu.cn (J.L.)
[2] Key Laboratory of Applied Geophysics, Ministry of Natural Resources of PRC, Changchun 130026, China
[3] Delaware State University, Dover, DE 19901, USA ; sxia@desu.edu
* Correspondence: zengzf@jlu.edu.cn

Received: 18 October 2018; Accepted: 27 November 2018; Published: 30 November 2018

Abstract: Lunar Penetrating Radar (LPR) is one of the important scientific systems onboard the Yutu lunar rover for the purpose of detecting the lunar regolith and the subsurface geologic structures of the lunar regolith, providing the opportunity to map the subsurface structure and vertical distribution of the lunar regolith with a high resolution. In this paper, in order to improve the capability of identifying response signals caused by discrete reflectors (such as meteorites, basalt debris, etc.) beneath the lunar surface, we propose a compressive sensing (CS)-based approach to estimate the amplitudes and time delays of the radar signals from LPR data. In this approach, the total-variation (TV) norm was used to estimate the signal parameters by a set of Fourier series coefficients. For this, we chose a nonconsecutive and random set of Fourier series coefficients to increase the resolution of the underlying target signal. After a numerical analysis of the performance of the CS algorithm, a complicated numerical example using a 2D lunar regolith model with clipped Gaussian random permittivity was established to verify the validity of the CS algorithm for LPR data. Finally, the compressive sensing-based approach was applied to process 500-MHz LPR data and reconstruct the target signal's amplitudes and time delays. In the resulting image, it is clear that the CS-based approach can improve the identification of the target's response signal in a complex lunar environment.

Keywords: lunar penetrating radar; lunar exploration; compressive sensing; lunar regolith modeling; signal processing

---

1. Introduction

The capability of ground-penetrating radar (GPR) to penetrate different materials makes it an effective and nondestructive geophysical tool for mapping the subsurface stratigraphy of the Moon to a given depth, which depends on the radar frequency and dielectric property of the lunar surface materials [1,2]. For example, the Lunar Radar Sounder (LRS) onboard Kaguya was used to detect the geological structure at depths of 4–5 km under the lunar surface [3,4]; the Apollo Lunar Sounder Experiment (ALSE) on the Apollo 17 spacecraft obtained a large amount of geological data from depths of 1–2 km below the surface of Moon [4,5]; and the dual-frequency Lunar Penetrating Radar (LPR) on the Yutu lunar rover, part of China's Chang'E-3 (CE-3) lunar mission, focuses on mapping the near-surface stratigraphic structure of the lunar regolith to a depth of several tens of meters [2,4,6–8].

The lunar regolith is formed by continuous meteoroid impacts on the lunar surface [9], resulting in the expulsion of surficial materials in the form of ejecta deposits, which are then comminuted, welded, overturned, mixed, altered, and homogenized by subsequent impacting and space weathering events

[6]. The composition and structure of the lunar regolith hold vital clues about the geology and impact history of the Moon. Those clues are critical for quantifying potential resources for future lunar exploration and determining the engineering constraints for human outposts [6,10–12]. The LPR of the CE-3 mission provides the opportunity to explore the subsurface structure and vertical distribution of the lunar regolith with a high resolution. LPR has two channels with center frequencies at 60 and 500 MHz [4]. Compared to the LRS (frequency of 5 MHz [3]) and ALSE(frequencies of 5, 15, and 150 MHz [5]), LPR can map the composition and structure of the regolith at shallower depths and with a higher range of resolution due to the higher frequencies used, especially the channel with a frequency at 500 MHz [6,13]. To date, many studies have focused on this 500-MHz LPR data. Fa et al. [6] and Lai et al. [2] estimated the near-surface structure with four major stratigraphic zones using the 500-MHz LPR data. Dong et al. [14] calculated the lunar surface regolith parameters in the CE-3 landing area, including its permittivity, density, conductivity, and FeO + TiO$_2$ content based on the 500-MHz LPR data. Feng et al. [15] derived the regolith's permittivity distribution laterally and vertically by processing the 500-MHz LPR data. In the analysis and evaluation of LPR data, the response signal caused by discrete reflectors beneath the lunar surface provides very useful information [4,6,7,14–16]. For example, the hyperbolic signatures produced by these targets are small with respect to radar wavelength, whose axes and vertices are functions of their position and relative dielectric characteristics [16,17]. In the lunar regolith, the most common subsurface materials are fine-grained regolith and basalt debris [1,4], and the layered reflection is not obvious [7]. Moreover, there is extensive clutter and noise in LPR data images [15], such as the coupling between antennas and the lunar surface, electromagnetic interference, etc. [4,14], and these can partially or totally hide or distort the response signal of discrete reflectors in the regolith [18]. Although many corrections have been applied to LPR data, such as background removal [14], amplitude compensation [14,15], band-pass filtering [13], and bi-dimensional empirical mode decomposition filtering [7], only a few reflections can be clearly identified [7,14,15]. Therefore, improving the capability to identify response signals of the discrete reflectors from LPR data is necessary.

The emerging compressive sensing (CS) theory maintains that sparse signals can be reconstructed from a small set of non-adaptive linear measurements by solving a convex problem with high probability [18–20]. The CS theory has already been used in radar signal denoising and imaging [18,21,22]. With basic information [19], CS has strong anti-interference ability. It can still perfectly reconstruct a radar signal if some elements are lost or polluted by noise [21]. In this paper, we propose a processing approach based on the CS theory to improve the capabilities of target signal extraction from the 500-MHz LPR data. First, we sparsed the LPR data in the frequency domain [23–25]. Then, we randomly selected a set of Fourier series in the proper frequency band to estimate the amplitudes and time delays using atomic norm minimization and total-variation (TV) norm minimization. In our approach, any one element in the LPR data is important, or rather, unimportant. This randomness greatly improves the target response signal extraction capability.

Numerical analyses of the CS algorithm's performance were performed, including denoising performance analysis, computational stability analysis, and computation complexity analysis. The result of a numerical example using a complex 2D lunar regolith model with clipped Gaussian random permittivity verifies the validity of the CS algorithm for LPR data. Finally, the compressive sensing-based approach was applied to estimate the signal amplitudes and time delays from the 500-MHz LPR data. The amplitude is the impedance contrast at the interface, or reflection coefficient, and the time delay indicates the detected target's depth. By studying the amplitudes and time delays, the position and shape of response signals caused by discrete reflectors beneath the lunar surface can be extracted.

This paper is structured as follows. In Section 2, an introduction to the compressive sensing-based approach and some numerical analyses are given. Section 3 describes a complex 2D lunar regolith numerical simulation model with clipped Gaussian random permittivity, which was established to verify the CS algorithm. In Section 4, we present our application of the compressive sensing-based approach to estimate the amplitudes and time delays from the 500-MHz LPR data. Finally, in Section 5, we draw some conclusions.

2. Methodology and Preliminary Numerical Tests

2.1. Signal model

The basic principle of GPR is the transmission of an electromagnetic (EM) radar pulse to image the subsurface targets or geological layers. The received signal $x(t)$ can be written as [19]:

$$x(t) = \sum_{j=1}^{L} a_j g(t - \tau_j),$$

(1)

where L is the number of reflected waves in the received signal. It should be larger than the number of scatterers and layers considering multiple reflection phenomena. $\{\tau_j\}_{j=1}^{L}$ is the total trip delay from the transmitting antenna to the target/layer j and back to the receiving antenna; $\{a_j\}_{j=1}^{L}$ is the amplitude, which is proportional to the target's radar cross-section (RCS), dispersion attenuation, and spreading losses throughout propagation; and $g(t)$ is the transmitted pulse. Thus, the received signal is essentially a time delay and a scaled version of the transmitted pulse. If the amplitudes and time delays are known, we can reconstruct the reflection signal.

2.2. CS Algorithm Using Random Fourier Series

Since the received signal $x(t)$ is confined to the interval $[0, \tau)$, we can extend $x(t)$ in a Fourier series as:

$$x(t) = \sum X[k] e^{i\frac{2\pi}{\tau}kt}, t \subset [0, \tau),$$

(2)

where:

$$X[k] = \frac{1}{\tau} \int_0^{\tau} x(t) e^{-i\frac{2\pi}{\tau}kt} dt,$$

(3)

Substituting Equation (1) into Equation (3), we get:

$$X[k] = \frac{1}{\tau} G\left(\frac{2\pi}{\tau}k\right) \sum_{j=1}^{L} a_j e^{-i\frac{2\pi}{\tau}k\tau_j}$$

(4)

where $G(\omega)$ denotes the continuous time Fourier transformation of $g(t)$.

For a set κ of K indices for which $G\left(\frac{2\pi}{\tau}k\right) \neq 0, \forall k \in \kappa$, such an integer subset exists for a UWB (ultra-wideband) radar transmitted pulse due to its very large relative bandwidth. Equation (4) can be rewritten as:

$$Y[k] = \frac{X[k]}{\frac{1}{\tau} G\left(\frac{2\pi}{\tau}k\right)} = \sum_{j=1}^{L} a_j e^{-i\frac{2\pi}{\tau}k\tau_j}$$

(5)

$V(t)$ denotes the $k \times L$ Vandermonde matrix given by $e^{-i\frac{2\pi}{\tau}k\tau_j}$, where $t = \{\tau_1, \cdots, \tau_L, \}$ is the vector of the unknown time delays. In addition, let $\alpha = \{a_1, \cdots, a_L\}^T$ and $y = \{Y_1, \cdots, Y_k\}^T$. The formulation of the relationship between the signal's Fourier series coefficients (y) and its unknown parameters (amplitudes and time delays) is obtained [24] as follows:

$$y = V(t)\alpha,$$

(6)

Given vector y, the total-variation (TV) norm is the continuous Fourier series version that is used to estimate the amplitudes and time delays [19,26], which can be interpreted as finding the shortest linear combination of elements taken from a continuous and infinite dictionary. The TV norm minimization problem in Equation (6) is expressed as:

$$min \, \|\alpha\|_{TV}$$
$$subject \; to \; \|V\alpha - y\|_2 \leq \delta \tag{7}$$

where δ is the noise level, and α can be recovered with a precision inversely proportional to δ [26].

This is a convex optimization problem that can be solved efficiently by many different algorithms. In this paper, in order to make the selected δ widely adaptable, we fixed it with the data noise level by the following formula:

$$\delta = \frac{\|y\|_2}{2N_y}. \tag{8}$$

where N_y is the length of vector y.

To acquire the Fourier series coefficients, we employed the Xampling scheme (described in the papers [19,25,27]), which enables the extraction of the necessary samples of Fourier series coefficients at a sub-Nyquist rate. Consecutive Fourier series coefficients can be easily obtained but, here, we used a nonconsecutive set of Fourier series coefficients that we randomly selected in a distributed manner from wide frequency aperture, which greatly increases the resolution of the underlying signal [28]. In other words, any Fourier series is important, or any Fourier series is not important. Tang et al. [28] proposed an atomic norm minimization approach, similar to the total-variation (TV) norm minimization, to recover the missing Fourier series coefficients. Assume that a subset of entries κ are selected at random and form the set $\{y_k, k \subset \kappa\}$ of consecutive Fourier coefficients, as prescribed in the paper [28]; then, a natural algorithm for estimating the missing samples from a sparse sum of complex exponentials is the atomic norm minimization problem:

$$min \, \|\tilde{y}\|_A$$
$$|\tilde{y}_j - y_j| < \delta, j \in \kappa, \tag{9}$$

where $\|\tilde{y}\|_A$ is the atomic norm of A associated with conv(A) (the convex hull of A), defined by:

$$\|\tilde{y}\|_A = inf \, \{t > 0 | \tilde{y} \in tconv\,(A)\}, \tag{10}$$

Equation (9) is equivalent to the following semi-definite program (SDF):

$$min \; trace\,(Toep\,(u)) + t$$
$$subject \; to \; \begin{bmatrix} Toep\,(u) & y \\ y^* & t \end{bmatrix} \geq 0 \tag{11}$$
$$|\tilde{y}_j - y_j| < \delta, j \in \kappa,$$

where $Toep\,(u)$ denotes the Toeplitz matrix, whose first column is equal to u; $u_k = \sum c_j \bar{c}_{j-k}$ and c are Fourier series coefficients. By solving Equation (11), the whole set $\{y_k, k \subset \kappa\}$ of consecutive Fourier series coefficients can be estimated. Then, the amplitudes and time delays can be estimated by the TV norm [19].

2.3. Numerical Analysis

In this subsection, we present the performance analysis of the estimated parameters of the LPR data using the CS algorithm. In order to get $x(t)$, one calculates the LPR response of a 1D simple lunar regolith model (Figure 1a) using the finite-difference time-domain (FDTD) algorithm [29]. The simulation parameter settings are consistent with CE-3 Channel 2 [30] (Table 1). The time step of the LPR is 0.3125 ns. Therefore, we need to sparse the simulation LPR data from 0.03125–0.3125 ns. The simulated amplitudes ($\{a_j\}_{j=1}^{L}$) and time delays ($\{\tau_j\}_{j=1}^{L}$) are listed in Table 2.

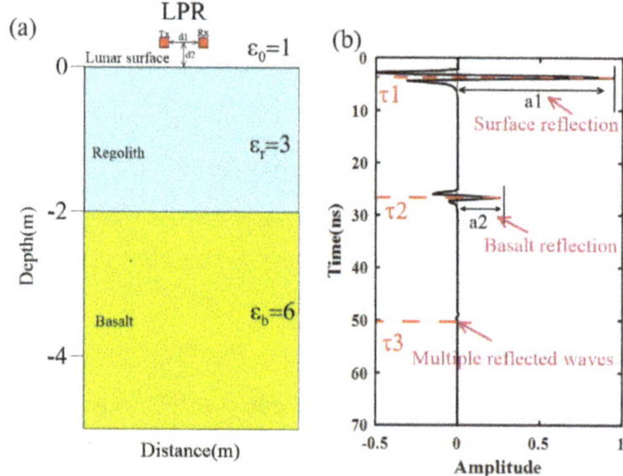

Figure 1. (a) The 1D lunar regolith model (Rx: receive antenna; Tx: transmitting antenna) and (b) the LPR response signal of the 1D lunar regolith model with the direct waves removed ($x(t)$).

Table 1. Simulation parameters.

Parameter	Value	Direction
Height of antenna	0.3 m	d2 in Figure 1a
Offset of Tx and Rx	0.32 m	d1 in Figure 1a
Transmitted Waveform	Ricker	$g(t)$
Center frequency	500 MHz	
Absorbing boundary	C-PML	
Thickness of absorbing boundary	0.1 m	10 PML layers
Discrete grid	0.01×0.01 m	the size of the grid cells
Time step	0.03125 ns	
Time window	70 ns	

Table 2. Amplitudes and time delays of $x(t)$.

Amplitude (a)	Value	Time Delay (τ)	Value (ns)
a1	0.9421	$\tau1$	3.7500
a2	0.2546	$\tau2$	26.5625
a3	-0.0092	$\tau3$	49.6875

Figure 2a is the continuous time Fourier transformation ($G(\omega)$) of a 500-MHz Ricker. Obviously, LPR is a kind of ultra-wideband radar (UWB) [4] and has a large bandwidth (>1 GHz). In [19], Xia et al. stated that one can reconstruct radar signals in a larger subset of entries κ. In this study, we limited the bandwidth to improve the operational efficiency and increase the signal resolution from the noise.

For the first numerical experiment, we chose three subsets (Bandwidth 1 (B1), B2, B3; Figure 2a) of κ and randomly selected 30 random Fourier series. The B1 bandwidth is from 400–600 MHz with $G\left(\frac{2\pi}{\tau}k\right) \neq 0$; the B2 bandwidth is from 800–1000 MHz with $G\left(\frac{2\pi}{\tau}k\right) \neq 0$; and the B3 bandwidth is from 1200–1400 MHz with $G\left(\frac{2\pi}{\tau}k\right) = 0$.

As this CS algorithm involves the random choice of parameters (Fourier coefficients), even if we set the same frequency band and number of Fourier coefficients, the estimated amplitudes and time delays will be different in each calculation (Figure 3). Therefore, in order to evaluate the performance of the CS algorithm accurately, for the final value in the numerical experiment, we used the average value obtained by executing the algorithm 60 times, selected by checking the convergence of its results. The standard deviation information is given to describe the stability of the algorithm. In Figure 3, the average value (AVG) is 0.2553 and the standard deviation (σ) is 0.0025. Obviously, this CS algorithm is stable. To account for the possibility of processing data collected over a larger spatial scale, we recorded the time cost of the numerical experiments. The parameters of the computer used to execute the numerical experiment are listed in Table 3.

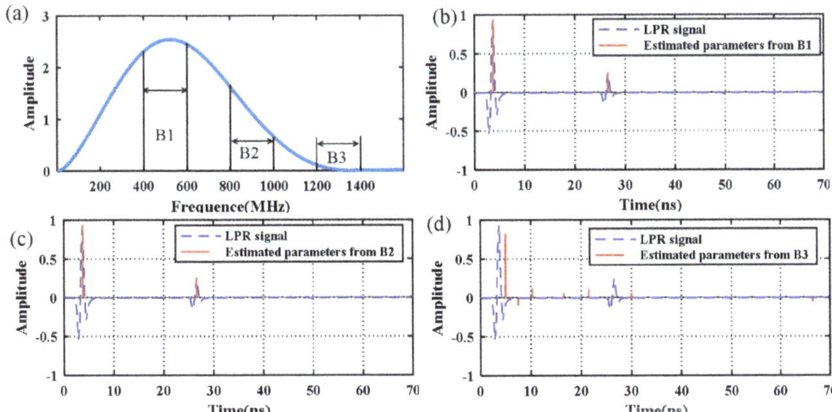

Figure 2. The results of the first numerical experiment. (**a**) The continuous time Fourier transformation of $g(t)$. (**b**) Estimated parameters by the CS algorithm in Bandwidth 1 (B1). (**c**) Estimated parameters by the CS algorithm in B2. (**d**) Estimated parameters by the CS algorithm in B3.

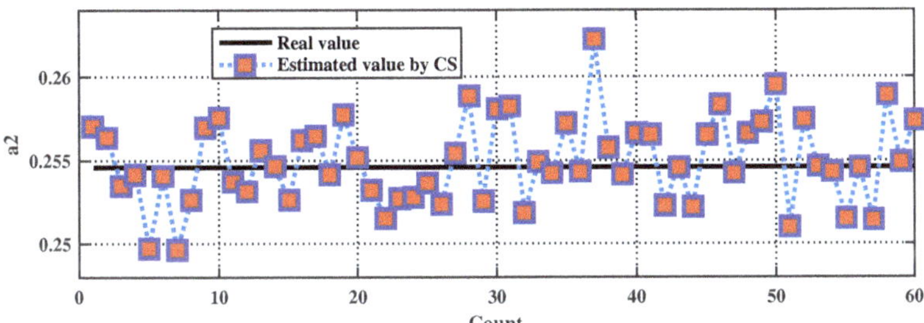

Figure 3. Estimated value of a2 from B1 obtained by executing the CS algorithm 60 times for numerical experiments. The average value (AVG) is 0.2553, and the standard deviation (σ) is 0.0025.

Table 3. The parameters of the computer used to execute the numerical experiment.

Type	Laptop
Operating system	Microsoft Windows 10
CPU	Intel(R) Core(TM) i7-6820HQ
RAM	16 GB

Figure 2 shows the results of parameter estimation in different bandwidth subsets. The estimated amplitudes $\{a_j\}_{j=1}^{L}$ and time delays $\{\tau_j\}_{j=1}^{L}$ are listed in Table 4. The relative error between the estimated value and the simulated value (Table 2) is defined as err_rel (%).

Table 4. Estimated parameter value statistics table of the first numerical experiment.

Frequency Band (MHz)	Amplitudes (a), Time Delays (τ)	Value	err_rel (%)	Standard Deviation (σ)	Time Cost (s)
B1[400–600]	a1	0.9421	0	0.0006	
	a2	0.2553	0.3	0.0025	
	a3	0	~	~	160
	τ1	3.7500 ns	0	0	
	τ2	26.5625 ns	0	0	
	τ3	~	~	~	
B2[800–1000]	a1	0.9420	0.1	0.004	
	a2	0.2570	1.0	0.0045	
	a3	0	~	~	159
	τ1	3.7500 ns	0	0	
	τ2	26.5625 ns	0	0	
	τ3	~	~	~	
B3[1200–1400]	a1	0.8325	11.63	0.2	
	a2	0	~	~	
	a3	0	~	~	160
	τ1	6.0 ns	60	3	
	τ2	~	~	~	
	τ3	~	~	~	

According to the results, both B1 and B2 can estimate amplitudes and time delays very well, and the amplitudes calculated by B1 have more accurate results. However, in B3, as mentioned in the theorysection, since the smaller value of the Fourier series coefficients and a part of $G\left(\frac{2\pi}{\tau}k\right)$ approximate to zero, the CS algorithm cannot accurately estimate the parameters. Considering that the center frequency of $g(t)$ (500 MHz) is included in B1 ([400–600] MHz), we recognize that the CS algorithm can obtain an accurate estimation of the signal parameters from the LPR data using a random Fourier series subset within a limited bandwidth around the center frequency. To gain a better understanding of the performance and reliability of the algorithm and to identify the best settings (the number of Fourier series coefficients and the bandwidth extension), we conducted a second and third numerical experiment with a focus on controlling the variables.

In the second numerical experiment, the number of Fourier series was fixed to 30, and the frequency bandwidths were set at 30% (B4[425–575] MHz), 50% (B5[375–625] MHz), and 60% (B6[350–650] MHz) of the central frequency, with centering on the central frequency (500 MHz). The results are listed in Table 5. From those results, the accuracy and reliability of the CS algorithm are satisfactory for the three frequency bands. When the frequency range is extended from B1–B5 MHz, there is a rather significant change of the algorithm performance: the error becomes higher and similar to the error obtained in B2. In order to achieve a better estimated accuracy over a larger bandwidth, the number of Fourier coefficients needs to be increased. In other words, the number of Fourier series needs to meet a certain density in the frequency band. However, the increase of Fourier series density reduces the randomness of the algorithm. In our CS algorithm, a random Fourier series can increase the resolution of the underlying signal in actual LPR data. We also noticed that, although a larger bandwidth can provide increased randomness in Fourier series selection, time costs and estimation errors increase: when the bandwidth increases from 150 to 300 MHz, the time cost increases by 2.5 times.

Table 5. Estimated parameter value statistics table of the second numerical experiment.

Frequency Band (MHz)	Amplitudes (a), Time Delays (τ)	Value	err_rel (%)	Standard Deviation (σ)	Time Cost (s)
B4[425–575]	a1	0.9421	0	0.0002	
	a2	0.2550	0.3	0.0015	
	a3	0	~	~	113
	τ1	3.7500 ns	0	0	
	τ2	26.5625 ns	0	0	
	τ3	~	~	~	
B5[375–625]	a1	0.9420	0.1	0.004	
	a2	0.2565	0.9	0.0036	
	a3	0	~	~	265
	τ1	3.7500 ns	0	0	
	τ2	26.5625 ns	0	0	
	τ3	~	~	~	
B6[350–650]	a1	0.9420	0.1	0.005	
	a2	0.2568	0.9	0.0062	
	a3	0	~	~	402
	τ1	3.7500 ns	0	0	
	τ2	26.5625 ns	0	0	
	τ3	~	~	~	

In the third numerical experiment, the frequency range was fixed to B1[400–600] MHz, and the number of Fourier series was set at 10 or 60. The results are listed in Table 6. In this numerical experiment, we cannot accurately estimate the amplitude and time delay using a random Fourier series subset with 10 coefficients. The CS algorithm requires a sufficient number of Fourier series coefficients to ensure reliability. By doubling the number of Fourier coefficients (from 30 to 60), a higher stability of the algorithm can be achieved (in terms of smaller standard deviation) at a limited additional computational cost (only 20% longer execution time). Although we decided to work with 30 coefficients in this paper, in some future scenarios, it may be useful to use more coefficients. However, an increase in the number of Fourier series, such as the 60 Fourier series in Table 6, means an increase in time cost and a decrease in randomness.

Therefore, considering the estimation error, randomness, and time cost, a random Fourier series subset with 30 coefficients chosen within 400–600 MHz is one of the best settings for 500-MHz LPR data processing. The CS algorithm can be executed on a personal computer. Single trace data can be processed in a few minutes. However, if, in some cases, we need to increase the frequency range and the number of Fourier series, the increase in time cost is inevitable. In addition, the time cost limitations of this CS algorithm are also possible when dealing with larger scale problems.

In order to assess the anti-interference ability and robustness of the CS algorithm, we designed the fourth and fifth numerical experiments with noise.

In the fourth numerical experiment, we added two sine interferences to $x(t)$ using Equation (12) (Figure 4).

$$x(t) = x(t) + sin(2\pi * M_1 * t) * 0.1 + sin(2\pi * M_2 * t) * 0.2 \tag{12}$$

where (M_1, M_2) is the frequency of the sine interference and was set at (200 MHz, 800 MHz) and (450 MHz, 550 MHz). The parameter values were estimated by the CS algorithm using 30 random Fourier series in B1(400–600 MHz). The results are listed in Table 7. The accuracy of the CS algorithm is satisfactory. Since the Fourier series in our CS algorithm is discontinuous and random, although 450 and 550 MHz are included in the frequency range (400~600 MHz), we can still accurately estimate the amplitude and time delays.

Table 6. Estimated parameter value statistics table of the third numerical experiment.

Number of Fourier Series	Amplitudes (a), Time Delays (τ)	Value	err_rel (%)	Standard Deviation (σ)	Time Cost (s)
10	a1	0.9323	1	0.1049	152
	a2	0.2050	20.1	0.0473	
	a3	0	$\sim$	$\sim$	
	τ1	3.7500 ns	0	0.01	
	τ2	26.5625 ns	0	0.01	
	τ3	$\sim$	$\sim$	$\sim$	
60	a1	0.9421	0	0.0001	192
	a2	0.2550	0.1	0.0002	
	a3	0	$\sim$	$\sim$	
	τ1	3.7500 ns	0	0	
	τ2	26.5625 ns	0	0	
	τ3	$\sim$	$\sim$	$\sim$	

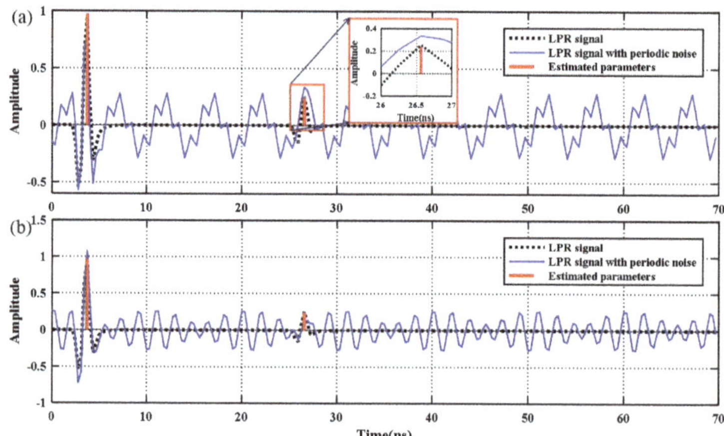

Figure 4. LPR data with periodic noise and parameters estimated by the CS algorithm. (**a**) The frequencies of the sine interference are 200 and 800 MHz; (**b**) the frequencies of the sine interference are 450 and 550 MHz).

Table 7. Estimated parameter value statistics table of the fourth numerical experiment.

Frequency of Sine Wave (MHz)	Amplitudes (a), Time Delays (τ)	Value	err_rel (%)	Standard Deviation (σ)	Time Cost (s)
200,800	a1	0.9749	3	0.001	162
	a2	0.2414	5	0.001	
	a3	0	$\sim$	$\sim$	
	τ1	3.7500 ns	0	0	
	τ2	26.5625 ns	0	0	
	τ3	$\sim$	$\sim$	$\sim$	
450,550	a1	0.98531	5	0.05	169
	a2	0.2358	7	0.03	
	a3	0	$\sim$	$\sim$	
	τ1	3.7500 ns	0	0	
	τ2	26.5625 ns	0	0	
	τ3	$\sim$	$\sim$	$\sim$	

In the fifth numerical experiment, we added -30, -20, and -10 dB of additive white Gaussian noise (AWGN) to $x(t)$ (Figure 5). The noise level is defined by the following formula:

$$AWGN\_dB = 10log_{10}\frac{P\_n}{P\_s},\tag{13}$$

where P_n is the noise signal power and P_s is the original signal power.

The parameter values were determined by the CS algorithm using 30 random Fourier series in B1(400–600 MHz). The results are listed in Table 8.

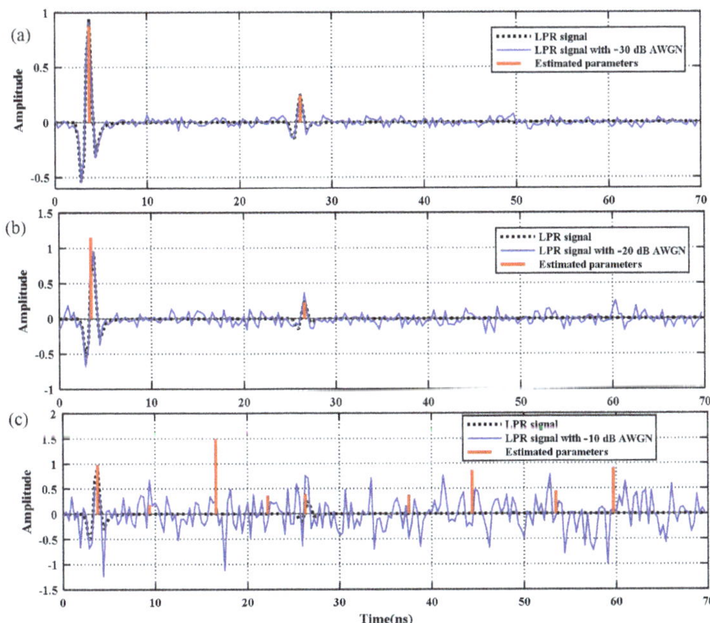

Figure 5. LPR data with Gaussian noise and the parameters estimated by the CS algorithm. (**a**) Noise level: −30 dB; (**b**) noise level: −20 dB; (**c**) noise level: −10 dB.

Table 8. Estimated parameter value statistics table of the fifth numerical experiment.

AWGN Noise Level (dB)	Amplitudes (a), Time Delays (τ)	Value	err_rel (%)	Standard Deviation (σ)	Time Cost (s)
30	a1	0.8802	7	0.002	
	a2	0.2450	0.3	0.003	
	a3	0	~	~	169
	τ1	3.7500 ns	0	0	
	τ2	26.5625 ns	0	0	
	τ3	~	~	~	
20	a1	1.125	20	0.05	
	a2	0.2315	9	0.06	
	a3	0	~	~	170
	τ1	3.2500 ns	12	0	
	τ2	26.5625 ns	0	0	
	τ3	~	~	~	
10	a1	1.022	0.8	0.2	
	a2	~	~	~	
	a3	~	~	~	165
	τ1	3.7500 ns	0	~	
	τ2	~	~	~	
	τ3	~	~	~	

Obviously, the signal parameters can be accurately estimated from signals with noise in general (Figure 5a,b). Due to the wide frequency range of the AWGN, when the noise level is 10 dB (Figure 5c), the CS algorithm cannot distinguish between the reflection signal and the false waveform generated by the AWGN. This requires researchers to apply their judgment in practical applications.

These numerical experiments indicate that the CS algorithm used with a random Fourier series can estimate a target's signal parameters from noisy LPR data. This CS algorithm can reconstruct the amplitudes and time delays with high efficiency and high precision under an appropriate bandwidth and Fourier series.

3. Algorithm Verification Using 2D the Random Regolith Model

In order to verify that the CS algorithm is effective for LPR data, we built a complex lunar regolith model (Figure 6) and calculated the LPR response using the 2D FDTD algorithm [29]. We set the numerical simulation parameters to be the same as those in the numerical experiment described in Section 2.3; the parameters are presented in Table 1 in Section 2. The LPR moving route was from 1–9 m in distance, and the sampling interval on the moving route was 0.05 m.

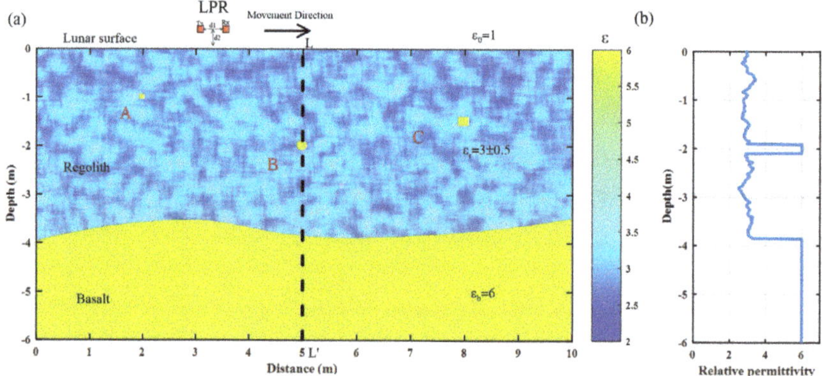

Figure 6. (a) The 2D random regolith model. (b) Permittivity trend at 5 m (black dotted line in (a)).

The model size was 10 m in the horizontal direction and 6 m in the vertical direction. To make the model structure similar to real lunar regolith, the following settings were applied.

(1)   The relative permittivity is proportional to the depth. The average relative permittivity of the lunar regolith is 3.0. Therefore, we improved the formula of the relationship between relative permittivity and depth [31,32] as follows:

$$\varepsilon_r = 1.919^{1.92\frac{z+12.2}{z+18}} + 0.4 \tag{14}$$

(2)   A Gaussian random field was used to model the regolith. A great deal of natural science data display marked Gaussian characteristics [33–35]. We built the lunar regolith relative permittivity model with clipped Gaussian random field theory [36]. A relative permittivity that is set from 2.5–3.5 satisfies the Gaussian random distribution.

(3)   The formation of lunar regolith indicates the existence of detritus [6]. In the LPR data, most reflections are formed by basalt debris [7]. Three strongly-reflecting debris materials (A: a square with a side length of 0.1 m; B: a circle with a radius of 0.1 m; C: a square with a side length of 0.2 m; Figure 6a) were incorporated into the 2D regolith model with the same relative permittivity as basalt ($\varepsilon_b = 6$).

Figure 7a is a snapshot of a radar wave in 28.8 ns at 5 m. The distortion of electromagnetic waves passing through random inhomogeneous media is clearly visible. Figure 7b shows the forward simulation result with direct waves removed. In the simulated LPR data image, the response of the strongly-reflecting debris is mixed with the response of weak reflectors in a random medium. The parameter values were estimated by the CS algorithm using 30 random Fourier series in B1 (400–600 MHz). Figure 8a is a single trace at 5 m (red line in Figure 7b) and the amplitudes and the time delays estimated by the CS algorithm. The CS algorithm not only accurately estimates the time delays ($\{\tau_j\}_{j=1}^{L}$) of a strong reflection response, but also restores the amplitudes ($\{a_j\}_{j=1}^{L}$). Figure 8b is the color image of absolute amplitudes ($|\{a_j\}_{j=1}^{L}|$). Compared to the original LPR image (Figure 7b), responsive recognition is greatly improved. According to the amplitude's absolute value, the reflected response signal can be located and identified. Moreover, some LPR responses of small debris in random inhomogeneous media are available. One can identify some complete weak responses based on the continuity of the amplitude to enrich the interpretation process.

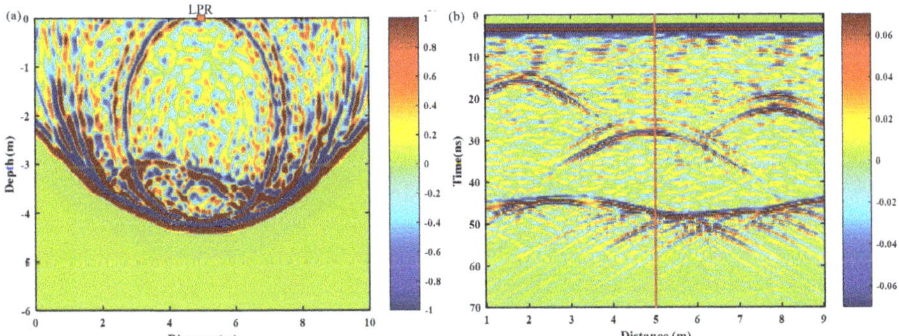

Figure 7. (**a**) A snapshot of a radar wave in 28.8 ns at 5 m. (**b**) Simulation results of the 2D regolith model.

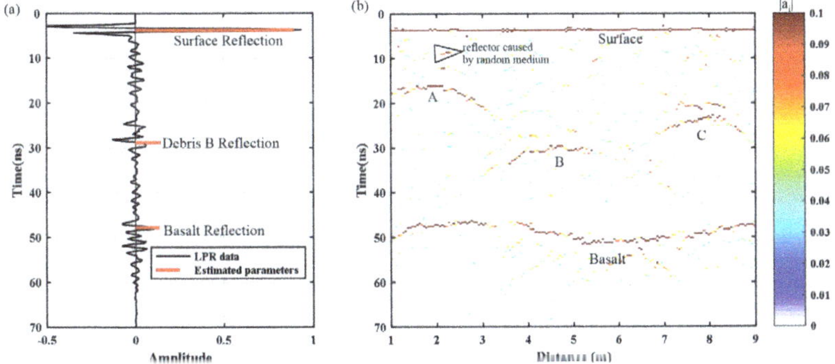

Figure 8. (**a**) A single trace in 5 m (red line in Figure 7b) and the parameters estimated by the CS algorithm. (**b**) Image of the estimated absolute amplitudes of the simulated LPR profile.

4. Processing and Analyzing LPR Data

4.1. Preprocessing

LPR is one of the important scientific systems onboard the Yutu lunar rover in the scope of China's Chang'E-3 lunar mission. The CE-3 landing site is in northern Mare Imbrium at 44.1213°N, 19.5115°W at an elevation of −2.627 km [6]. The Yutu lunar rover's route is shown in Figure 9. The raw LPR data are archived and distributed by the National Astronomical Observatories, Chinese Academy of Sciences (NAOC) (http://moon.bao.ac.cn/) [13]. In this paper, we focus on the data collected by the 500-MHz Level 2C data (a band-pass filter based on fast Fourier transform (FFT) filtering was done [7,13]) from C to D (Figure 9). The distance of the CD route is 10 m.Considering the unique characteristics of LPR data acquisition, we performed a longitudinal displacement correction, removing the repeating paths and velocity interpolation to extract the LPR data (Figure 10a) from the raw data [7]. Although the LPR data were successfully extracted, it is difficult to strip the direct waves from the original data. According to calibration tests that were performed by the Institute of Electronics, Chinese Academy of Sciences, the time delay of the radar echo signal transmitted from the lunar surface is 28.203 ns for Channel 2; thus, the lunar surface is characterized by a 28.203-ns time delay correction [30]. Then, we applied an automatic gain control (AGC) method to recover the amplitude loss from spreading and scattering [2] and removed the background using a median filter, as is usually used for GPR data for Earth-based applications. Since most of the effective information of the 500-MHz LPR data are concentrated within 100 ns [6,7], we only processed the data between 0 and 100 ns.

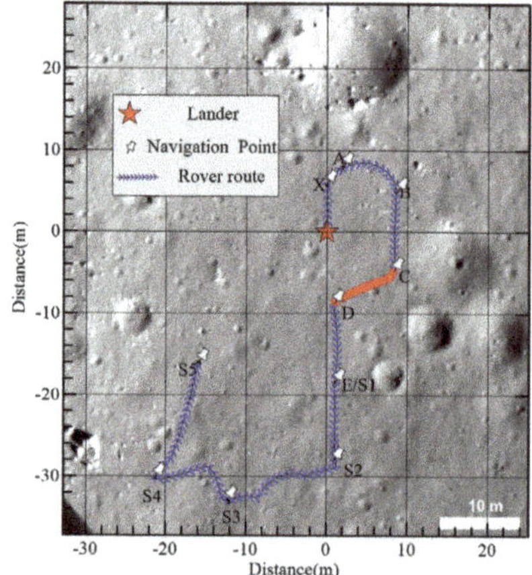

Figure 9. Yutu lunar rover's route [6,16]. The red line from C to D is the research route (10 m). We added a distance coordinate system, with the lander as the coordinate origin (0, 0).

4.2. Parameter Estimation Using the CS Algorithm

Based on the numerical experiments' results in the theoretical part of this study, we randomly selected 30 Fourier series coefficients from 200–600 MHz to estimate the amplitudes and time delays using the CS algorithm. Figure 11 is a single trace of the LPR data and the parameters estimated by the CS-based approach. From Figure 11, we can see that the estimated parameters are quite consistent with the preprocessed LPR data (Figure 11b). The aliased waveform image was converted to clear discrete reflection amplitudes and time delays. Figure 12 shows the CD LPR profile data overlaid with the processing results. In Figure 12, due to the time delay correction and the removal of the background, the surface reflection signal is missing. Time delays were distributed between 10 and 70 ns. Below 70 ns, although the sporadic reflected signals could be identified, the amplitude was smaller and discontinuous. In order to highlight the response of a reflector in the regolith, we provide a color image of the absolute amplitude (Figure 13) with a threshold ($| \{a_j\}_{j=1}^{L} | > 1.5$). Compared with the continuous waveforms in raw data, the CS algorithm more easily locates and extracts reflections caused by buried targets. By analyzing the continuity of the time delays and the values of the amplitudes, we can reconstruct and identify response signals caused by discrete reflectors beneath the lunar surface (Figure 14).

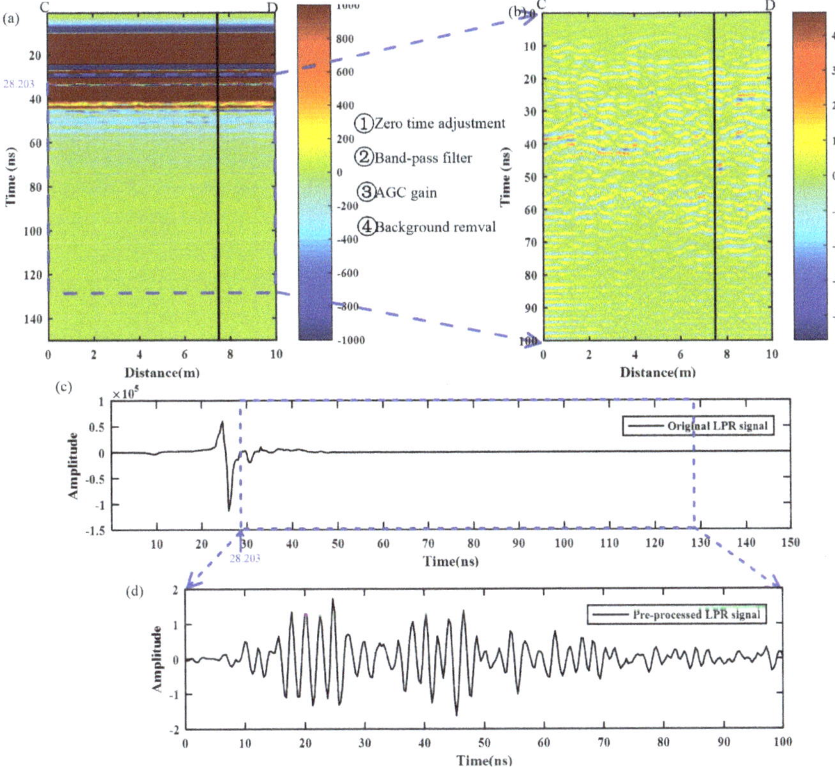

Figure 10. (**a**) The original LPR data extracted from the raw data, (**b**) the preprocessed LPR data, (**c**) a signal trace of the LPR data at 7.5 m (black line), and (**d**) a signal trace of the preprocessed LPR data at 7.5 m.

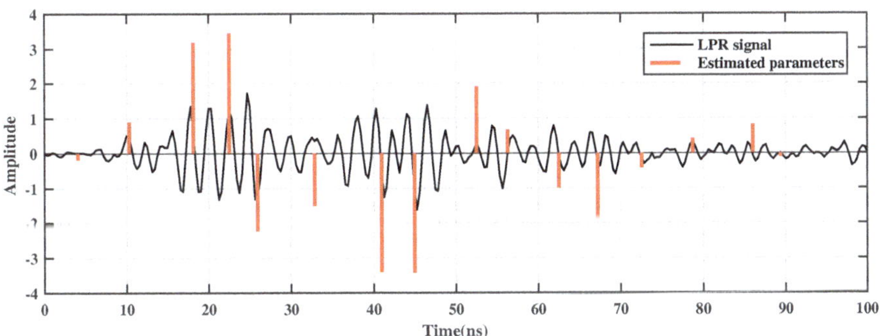

Figure 11. A single trace of LPR data at 6 m and parameters estimated by the CS-based approach.

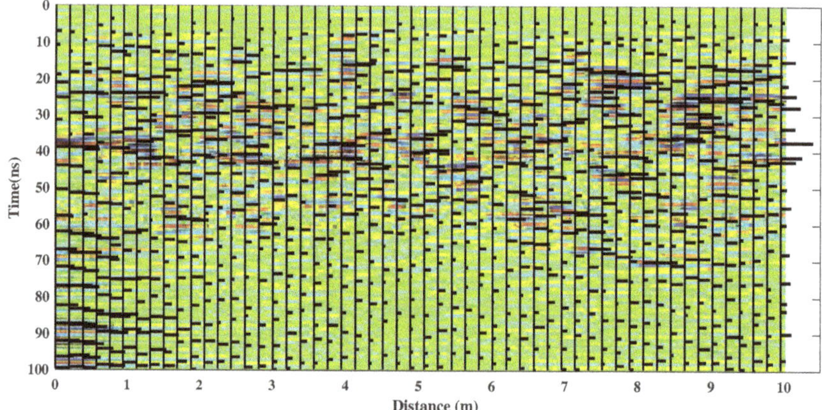

Figure 12. The LPR profile overlaid with the CS-based approach processing results.

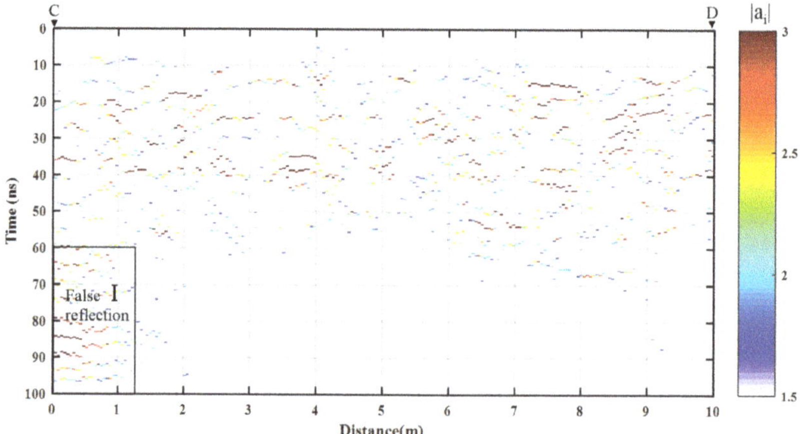

Figure 13. Absolute amplitude image of LPR data.

4.3. Results

When combining the forming mechanism of the regolith in the study area with the result of the CS-based approach, a further interpretation of the LPR data was revealed (Figure 14). A dielectric permittivity of 3.0 + i0.03 was used to calculate the depth. In the LPR data image, a three-layer structure could be clearly delimited as follows.

- The top layer (depth < 1 m) cannot extract reflection parameters at all. This is the fine-grained regolith part of the lunar regolith. In [6], this part was interpreted as a reworked zone. Even though the fine-grained regolith was composed of numerous layers, the layer thickness was typically on the order of several centimeters [37], which is much smaller than the LPR range resolution [4]. Therefore, it is difficult to extract the reflected signal from this layer.
- The middle layer (buried from 1–7 m) had the most signal reflectors beneath the lunar surface. It is the paleoregolith of mare basalt with much debris. After the processing step in the CS-based approach, the reflective response signal, which is difficult to extract from the original LPR data, became clear (Figure 13). The size of the reflection curve is often proportional to the volume of the debris [7]. On the basis of the continuity and the absolute value of the estimated amplitude,

some reflection response signals are marked by a red curve in Figure 14. Obviously, this is a good improvement for evaluating the LPR data.

- The last part is the basalt base. Since there is no obvious reflective interface between the regolith and mare basalt, one can distinguish this region by the distribution of a strong reflection response signal. Obviously, from Figure 13, there were no reflection signals ($| \{a_j\}_{j=1}^L | > 1.5$). The estimated reflection response at [I] (Figure 13) is a false reflection, considering the unusually persistent periodicity of this part of the LPR data.

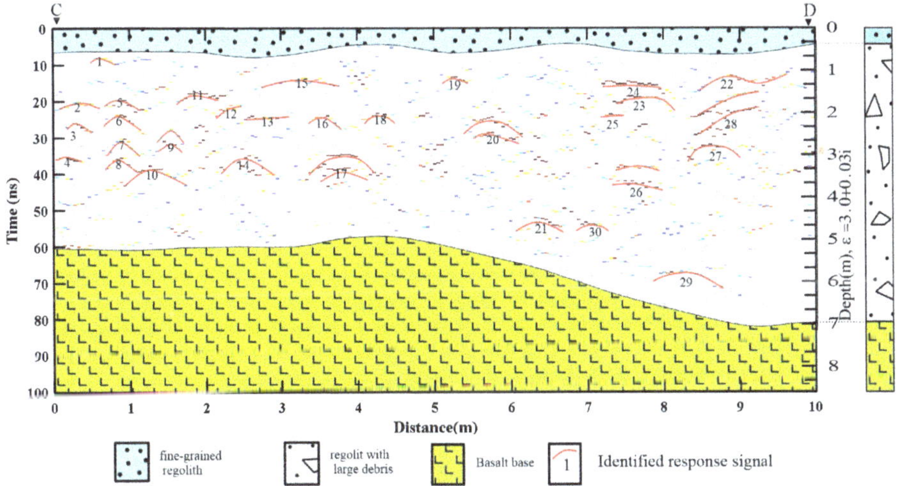

Figure 14. Interpretation of the LPR data from C to D (Figure 9).

4.4. Discussion

In the 2D random regolith model, a strong interface between the regolith and basalt is present. However, in the measured 500-MHz LPR data, there is no obvious reflective interface between the regolith and mare basalt. It is possible that there is a smooth variation in the permittivity on the interface. The overall reflection generated by a gradual interface has a different shape than the transmitted pulse [38], whereas a sudden permittivity change generates a reflection having a different amplitude, but the same shape as the transmitted pulse. This is a challenge for the CS algorithm. We hope to study this scenario in future work.

When the CS-based approach is applied to LPR data, there is an improvement in classifying the stratigraphic structure of the lunar regolith. The stratification result is consistent with previous studies [6,7,16], and we located the debris in the regolith. Moreover, the improvement in LPR data interpretation is not limited to this. It can provide a relatively accurate initial model for standard focusing methods (migration, beamforming, diffraction tomography, etc.). Accurate amplitudes and time delays also can significantly refine the accuracy of estimated lunar regolith parameters, such as permittivity and iron–titanium content [2,16]. It is critical to quantify potential resources for lunar exploration and for the engineering of human outposts.

5. Conclusions

In this paper, we propose a compressive sensing-based approach to reconstruct the amplitudes and the time delays of radar data. Numerical analysis of the CS-based approach performance was conducted and is herein discussed. The simulation results show that the approach is effective and stable. Then, successful estimations of the amplitudes and the time delays of the 500-MHz LPR data using the compressive sensing-based approach were achieved. The final result shows that the

CS-based approach can improve the capability of extracting a target's response signal from a complex lunar environment. In addition, these results provide valuable information about the lunar regolith structure and estimated parameters, such as the electrical parameters and the iron–titaniumcontent of the regolith.

Author Contributions: K.W. and Z.Z. wrote the article; Z.Z. supervised the project; L.Z. performed the pretreatment of the lunar data; K.W. and S.X. improved the reconstruction algorithms and procedures; K.W., Z.Z. and J.L. processed and analyzed the LPR data.

Acknowledgments: This research was supported by the Natural Science Foundation of China under Grant 41574097 and Grant 41504083.

Conflicts of Interest: The authors declare no conflict of interest.

Abbreviations

The following abbreviations are used in this manuscript:

LPR	lunar penetrating radar
CS	compressive sensing
GPR	ground-penetrating radar
LRS	Lunar Radar Sounder
ALSE	Apollo Lunar Sounder Experiment
CE-3	Chang'E-3
UWB	ultra-wideband
NAOC	National Astronomical Observatories, Chinese Academy of Sciences
AGC	automatic gain control
TV	total-variation
SDF	semidefinite program
FDTD	finite-difference time-domain
AWGN	additive white Gaussian noise

References

1. Russell, P.S.; Grant, J.A.; Williams, K.K.; Carter, L.M.; Garry, W.B.; Daubar, I.J. Ground penetrating radar geologic field studies of the ejecta of barringer meteorite crater, arizona, as a planetary analog. *J. Geophys. Res. Planets* **2013**, *118*, 1915–1933. [CrossRef]
2. Lai, J.; Xu, Y.; Zhang, X.; Tang, Z. Structural analysis of lunar subsurface with Chang'E-3 lunar penetrating radar. *Planet. Space Sci.* **2016**, *120*, 96–120. [CrossRef]
3. One, T.; Kumamoto, A.; Kasahara, K.; Yamaguchi, Y.; Yamaji, A.; Kobayashi, T.; Oshigami, S.; Nakagawa, H.; Goto, Y.; Hashimoto, K.; et al. The Lunar Radar Sounder (LRS) Onboard the KAGUYA (SELENE) Spacecraft. *Space Sci. Rev.* **2010**, *154*, 145–192. [CrossRef]
4. Fang, G.; Zhou, B.; Ji, Y.; Zhang, Q.; Shen, S.; Li, Y. Lunar penetrating radar onboard the Chang'E-3 mission. *Res. Astrono. Astrophys.* **2014**, *14*, 1607–1622. [CrossRef]
5. Porcello, L.J.; Jordan, R.L.; Zelenka, J.S. The Apollo lunar sounder radar system. *Proc. IEEE* **1974**, *62*, 769–788. [CrossRef]
6. Fa, W.; Zhu, M.; Liu, T.; Plescia, J.B. Regolith stratigraphy at the chang'e-3 landing site as seen by lunar penetrating radar. *Geophys. Res. Lett.* **2016**, *42*, 10179–10187. [CrossRef]
7. Zhang, L.; Zeng, Z.; Li, J.; Lin, J.; Hu, Y.; Wang, X. Simulation of the lunar regolith and lunar-penetrating radar data processing. *IEEE J. Sel. Top. Appl. Earth Obs. Remote. Sens.* **2018**, *11*, 655–663. [CrossRef]
8. Zhang, J.; Yang, W.; Hu, S.; Lin, Y.; Fang, G.; Li, C. Volcanic history of the imbrium basin: A close-up view from the lunar rover Yutu. *Proc. Natl. Acad. Sci. USA* **2015**, *112*, 5342–5347. [CrossRef] [PubMed]
9. Papike, J.J.; Simon, S.B.; Laul, J.C. The lunar regolith: Chemistry, mineralogy, and petrology. *Rev. Geophys.* **1982**, *20*, 761–826. [CrossRef]
10. Fa, W.; Wieczorek, M.A. Regolith thickness over the lunar nearside: results from earth-based 70-cm arecibo radar Obs. *Icarus* **2012**, *218*, 771–787. [CrossRef]

11. Papike, J.J.; Simon, S.B.; Laul, J.C. The lunar regolith: chemistry, mineralogy, and petrology. *Rev. Geophys.* **1982**, *20*, 761–826. [CrossRef]

12. Wilcox, B.B.; Robinson, M.S.; Thomas, P.C.; Hawke, B.R. Constraints on the depth and variability of the lunar regolith. *Meteorit. Planet. Sci.* **2015**, *40*, 695–710. [CrossRef]

13. Sun, Y.; Fang, G.Y.; Feng, J.; Xing, S.; Jing, Y.; Zhou, B. Data processing and initial results of Chang'E-3 lunar penetrating radar. *Res. Astron. Astrophys.* **2014**, *14*, 1623–1632.

14. Dong, Z.; Fang, G.; Ji, Y.; Gao, Y.; Wu, C.; Zhang, X. Parameters and structure of lunar regolith in chang'e-3 landing area from lunar penetrating radar (lpr) data. *Icarus* **2016**, *282*, 40–46. [CrossRef]

15. Feng, J.; Su, Y.; Ding, C.;Xing, S.; Dai, S.; Zou, Y. Dielectric properties estimation of the lunar regolith at CE-3 landing site using lunar penetrating radar data. *Icarus* **2017**, *284*, 424–430. [CrossRef]

16. Zhang, L.; Zeng, Z.; Li, L.; Huang, L.; Huo, Z.; Wang, K.; Zhang, J. Parameter Estimation of Lunar Regolith from Lunar Penetrating Radar Data. *Sensors* **2018**, *18*, 2907. [CrossRef] [PubMed]

17. Delbo, S.; Gamba, P.; Roccato, D. A fuzzy shell clustering approach to recognize hyperbolic signatures in subsurface radar images. *IEEE Trans. Geosci. Remote. Sens.* **2002**, *38*, 1447–1451. [CrossRef]

18. Ambrosanio, M.; Pascazio, V. A compressive-sensing-based approach for the detection and characterization of buried objects. *IEEE J. Sel. Top. Appl. Earth Obs. Remote. Sens.* **2017**, *8*, 3386–3395. [CrossRef]

19. Xia, S.; Liu, Y.; Sichina, J.; Liu, F. A compressive sensing signal detection for uwb radar. *Prog. Electromagn. Res.* **2013**, *141*, 479–495. [CrossRef]

20. Gurbuz, A.C.; McClellan, J.H.; Scott, W.R., Jr. Compressive sensing for subsurface imaging using ground penetrating radar. *Signal Process.* **2009**, *89*, 1959–1972. [CrossRef]

21. Zhu, L.; Zhu, Y.; Mao, H.; Gu, M. A New Method for Sparse Signal Denoising Based on Compressed Sensing. *Second. Int. Symp. Knowl. Acquis. Model.* **2009**, *1*, 35–38.

22. Metzler, C.A.; Maleki, A.; Baraniuk, R.G. From denoising to compressed sensing. *IEEE Trans. Inf. Theory* **2016**, *62*, 5117–5144. [CrossRef]

23. Maravic, I.; Vetterli, M. Sampling and reconstruction of signals with finite rate of innovation in the presence of noise. *IEEE Trans. Signal Process.* **2005**, *53*, 2788–2805. [CrossRef]

24. Vetterli, M.; Marziliano, P.; Blu, T. Sampling signals with finite rate of innovation. *IEEE Trans. Signal Process.* **2002**, *50*, 1417–1428. [CrossRef]

25. Michaeli, T.; Eldar, Y.C. Xampling at the rate of innovation. *IEEE Trans. Signal Process.* **2012**, *60*, 1121–1133. [CrossRef]

26. Candès, E.; Fernandez-Granda, C. Towards a mathematical theory of superresolution. *Commun. Pure Appl. Math.* **2013**, *67*,906–956. [CrossRef]

27. Gedalyahu, K.; Tur, R.; Eldar, Y. Multichannel sampling of pulse streams at the rate of innovation. *IEEE Trans. Signal Process.* **2011**, *59*,1491–1504. [CrossRef]

28. Tang, G.; Bhaskar, B.N.; Shah, P.; Recht, B. Compressed sensing off the grid. *IEEE Trans. Inf. Theory* **2013**, *59*, 7465–7490. [CrossRef]

29. Irving, J.; Knight, R. Numerical modeling of ground-penetrating radar in 2-d using matlab. *Comput. Geosci.* **2006**, *32*, 1247–1258. [CrossRef]

30. Zhou, N.; Zhou, P.; Yang, K.; Yuan, Y.; Gou, S. The preliminary processing and analysis of LPR channel-2b data from chang'e 3. *Sci. China Phys. Mech. Astron.* **2014**, *57*, 2346–2353. [CrossRef]

31. Dollfus, A. Physical properties of the lunar surface. *Catal. Ind.* **2014**, *6*, 114–121.

32. Kobayashi, T.; Kim, J.H.; Lee, S.R.; Araki, H.; Ono, T. Simultaneous observation of lunar radar sounder and laser altimeter of kaguya for lunar regolith layer thickness estimate. *IEEE GeoSci. Remote. Sens. Lett.* **2010**, *7*, 435–439. [CrossRef]

33. Haran, M. Gaussian random field models for spatial data. *Handb. Markov Chain. Mt-Carlo* **2009**, 449–478.

34. Jiang, Z.; Zeng, Z.; Li, J.; Liu, F.; Li, W. Simulation and analysis of GPR signal based on stochastic media model with an ellipsoidal autocorrelation function. *J. Appl. Geophys.* **2013**, *99*, 91–97. [CrossRef]

35. Zeng, Z.; Chen, X.; Li, J.; Chen, L; Lu, Q.; Liu, F. Recursive impedance inversion of ground-penetrating radar data in stochastic media. *Appl. Geophys.* **2015**, *12*, 615–625. [CrossRef]

36. Li, J.; Zeng, Z.; Liu, C.; Huai, N.; Wang, K. A study on lunar regolith quantitative random model and lunar penetrating radar parameter inversion. *IEEE GeoSci. Remote. Sens. Lett.* **2017**, *14*, 1953–1957.

[CrossRef]

37. Mitchell, J.; Carrier, W.; Costes, N.C.; Houston, W.; Scott, R. Surface Soil Variability and Stratigraphy at the Apollo 16 Site. *Geochim. Cosmochim. Acta* **1973**, *3*, 2437–2445.

38. Prokopovich, I.; Popov, A.; Pajewski, L.; Marciniak, M. Application of Coupled-Wave Wentzel-Kramers-Brillouin Approximation to Ground Penetrating Radar. *Remote. Sens.* **2017**, *10*, 22. [CrossRef]

MDPI

St. Alban-Anlage 66

4052 Basel

Switzerland

Tel. +41 61 683 77 34

Fax +41 61 302 89 18

www.mdpi.com

Remote Sensing Editorial Office

E-mail: remotesensing@mdpi.com

www.mdpi.com/journal/remotesensing